LOCAL HEROES
DO-IT-YOURSELF
SCIENCE

LOCAL HEROES
DO-IT-YOURSELF
SCIENCE

ADAM HART-DAVIS & PAUL BADER

Acknowledgements
We'd like to take this opportunity of thanking the
prop-builders who have been part of the team at some
stage — Chris Hill, John Francas, Paul King, Marty
Jopson and Jonathan Sanderson – our editor Lara
Speicher, our meticulous copy-editor Christine King,
and Jason Chapman who drew the splendid diagrams.

This book is published to accompany the TV series
Local Heroes. This series is produced for BBC
Television by Screenhouse Productions Ltd.
Series producer: Paul Bader

Published by BBC Worldwide Ltd,
Woodlands, 80 Wood Lane,
London W12 0TT

First published 2000
Copyright © Adam Hart-Davis and Paul Bader 2000
The moral right of the authors has been asserted.

ISBN 0 563 55165 8

Project editor: Lara Speicher
Copy-editor: Christine King
Designer: John Calvert
Illustrator: Jason Chapman

Printed and bound in Great Britain by
Butler & Tanner Ltd, Frome and London
Cover printed by Belmont Press Ltd,
Northampton

Contents

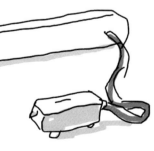

INTRODUCTION

Welcome to our world of hands-on science and technology. We started making *Local Heroes* television programmes in 1990, and from the very beginning we began to recreate the work of our heroes in the form of models. We found that by trying to do what they had done, we could better understand their ideas and their theories. By building models to explain the heroes' work, we were in some way able to share their triumphs.

To begin with, we had no money and no assistance, so we built all the machines ourselves, in our basement workshops, using any pieces of scrap material that were lying around. However, when we put these demonstrations on television, the viewers seemed to like their crudeness and simplicity, and so they remain crude and simple to this day. (Mind you, there's crude and crude: a few of the projects here do demand a certain level of dexterity and precision for best results.)

In this book you will find more than fifty of our favourite projects. Most of these are illustrations of technological inventions or of scientific ideas. For example, Faraday's candles and Faraday's electric motor are elegant demonstrations of what he did and, even though we know what will happen, the process of carrying out the demonstration helps us to think about how the science works.

However, some of the projects we describe in this book are pieces of apparatus you can build and then use to do genuine experiments, and find out answers that we don't know. Galileo described experiments as 'questions that the scientist asks of God'; a good scientist will ask good questions, and the answers should lead closer to the truth about Nature and our universe. You can do your own experiments with the Megaphones, the Whirling-arm machine, or Jelly-tower building.

Many of the things we built did not work well at first. Often we found that parts needed to be longer or stronger or wider or softer, and sometimes we had to go right back to the drawing board and start again. The methods and materials we describe in this book are the ones that did work for us in the end, and we hope that we have saved you some of the disappointments and disasters that happened to us. However, some of these projects are really tricky, and need skill and patience; don't be surprised if you don't get everything right first time! We have graded each one with stars:

RATING: ✳ EASY!

RATING: ✳✳ MODERATE

RATING: ✳✳✳ CHALLENGING

RATING: ✳✳✳✳ DIFFICULT

RATING: ✳✳✳✳✳ VERY DIFFICULT! (IN A CLASS OF ITS OWN)

We suggest that if you're not an experienced model-maker you should start with an easy one-star project and build up.

We've had a lot of fun building all these things, and learned a good deal in the process. We've found that just making and using them gives us a better understanding of scientific principles and technological tricks – and an insight into the people who did them first. We hope you too will enjoy following in the footsteps of the local heroes.

So good luck with all your ventures! We'd love to hear of your results – you'll notice that here and there we specifically ask for feedback. Do send your comments and suggestions to us at :

Local Heroes DIY Science
Screenhouse Productions
The Old Baptist School
378 Meanwood Road
Leeds LS7 2JF

e-mail: diyscience@screenhouse.co.uk

GUIDE TO MATERIALS

Each project has a list of what you'll need, but often this is just a guide — never think you can't do something because you don't have the exact item mentioned. There's usually room to improvise.

While we make the most of the junk we have lying around, sometimes we do have to go to a DIY or hardware shop and stock up. We don't have fixed rules about what we use, but for example when we say 'screws' to hold wood together, we probably mean 50mm No 6 posidrive countersunk screws, because they are easy to get hold of and to use, and they generally do the job. In this book we just say 'screws', and leave it to you to choose a suitable size — we don't want you to have to trail around trying to match exactly something we used only because it was there!

Where tools are concerned, we assume you'll have access to the common range of hammers, drills, screwdrivers, saws and so on. If a special tool is needed, then we'll spell it out.

Here are the sorts of materials we use:

Wood (natural and synthetic)
- Softwood: often 20 x 30mm or 20 x 40mm or, say, 20 x 300mm if we need planks.
- Dowel for shafts: this used to be sold as 1-inch dowel, but these days it's usually 22mm or 28mm (either of these sizes will be fine).
- MDF (medium-density fibreboard): 15mm to 30mm thick for heavy-duty bases or walls.

Plastic
- Sheets of flyweight envelope stiffeners are excellent for rigid strength and low weight. You can buy these from stationery shops.
- Containers: fizzy-drink bottles and milk bottles of various sizes are often useful. Also those small canisters for films, and screw-top plastic jars of cashews and other nuts.

Fastenings
- Screws: often 50mm No 6 posidrive countersunk.
- Nails: often 40mm lost-head (i.e. the head can easily be hammered flush with the wood for a smooth finish).
- Staples: either wire or plastic-padded, used to hold electrical or phone wires.
- Bolts: often 4mm or 6mm diameter. You can also get a length of 'threaded rod' in DIY stores – essentially a long headless bolt that's cut to length (it's very useful).

Glue
- For wood: white PVA glue is safe.
- For rubber or plastic: superglue (take special care).
- For metal, plastic, glass, etc: epoxy resin (e.g. Araldite).
- For paper: UHU or similar.
- We also sometimes use a hot-glue gun, an electrical gadget that melts glue for spreading – but it can get tricky.

Tape
- Gaffer tape or duck tape is generally 50mm wide, very sticky, and easy to tear across; obtainable in DIY or hardware shops. Carpet tape is similar.
- Parcel tape (brown plastic, 50mm) is very sticky and strong.
- Insulating tape (20mm, various colours) is easy to use but less strong and sticky.
- Waterproof tape is useful.

Metal frames
- Dexion (like giant Meccano) or equivalent, plus nuts and bolts – useful for strong frames.

Pipes
- Plastic plumbing pipes, drainpipes and sewage pipes are all useful. They are push-to-fit, easily cut with a hacksaw, and have a range of fittings available from DIY chains.
- Cardboard tubes are obtainable from stationery and art shops.
- Metal pipe can be borrowed from a vacuum cleaner.
- Flexible clear plastic pipe (PVC) comes in various sizes, from DIY or home-brew shops.

Metric/imperial measurements
We've used metric measurements, but for those who prefer to use inches, the basic equivalents are:

1cm = ½ in
2.5cm = 1in
5cm = 2in
10cm = 4in
20cm = 8in
30cm = 12in
1m = 3¼ feet

(These equivalents are rounded up or down – they don't have to be exact – but stick to one or other system.)

SAFETY

Don't lose your head – or even a couple of fingers!

Most people don't bother to read safety warnings, but please read this one. We would like you to be able to carry out some of the projects in this book, but we don't want anyone to be injured. If you're not a skilled craft worker, please get someone more experienced to help you, at least to begin with.

Specific safety advice is included with each project when appropriate.

Fire

Several of the projects call for the use of matches, lighters, candles and other naked flames. Please clear away any loose pieces of paper or other combustible material before starting one of these. Preferably do the project on a metal tray, or a kitchen worktop, or outside on a non-combustible surface. If possible have a fire extinguisher ready, or at least a big jug of water (provided no electricity is involved).

If you burn your finger over a match or candle, hold the burned finger under cold running water for a minute or two.

Steam heat

Steam looks pretty, but is dangerous because it carries a lot of heat, and can scald you badly. NEVER heat a closed container of water, which is highly dangerous; make sure there is at least one hole for the steam to escape, and keep away from the steam.

Sharp knives

For many of the projects you will need to cut up cardboard, plastic or other material with a sharp knife. We recommend craft knives or Stanley knives. Do always try to cut away from yourself, so that if the knife slips it does not go into you.

If you do cut yourself, wash the wound with lots of cold water, and tie it up with sticking plaster or a clean handkerchief. If it does not stop bleeding, go and get medical help.

Saws and drills

You will need to use such basic tools as saws and drills for many of these projects. They are sharp and potentially dangerous. Always use them on a good solid bench with good bright lighting, keeping your fingers away from the sharp blades. If you lack experience, get someone to show you how to use these tools.

Hammers

If you have had no practice, and you find it difficult to hammer in a nail without hammering your fingers, try holding the nail with a clothes peg or poked through a piece of cardboard, so that your fingers are at least 5cm away from the nail.

Glue

When you use superglue, be careful not to get it on your skin. Read the warnings and instructions on the tube. Other glues are less dangerous, but we strongly advise you to do your gluing on an old newspaper on a table; so that if it spreads it won't do any damage.

Electricity and power tools

Mains electricity is part of only one demonstration (the Fax Machine), and you should tackle this only if you are confident you can connect circuits safely. But many of the others might involve power tools in their construction. Keep them well away from any water involved, and read the safety instructions first.

Protective clothing

We wear work-gloves to protect our hands from heat or sharp edges, goggles to protect eyes from steam, dust and fumes, and masks when sanding, sawing or dealing with vapour. These items are available cheaply from DIY stores, and we encourage you to use them.

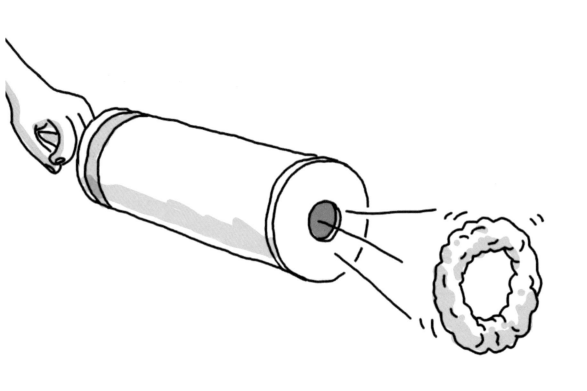

CHAPTER ONE
VAPOUR, SMOKE & STEAM

HOVERCRAFT

Sir Christopher Cockerell's brilliant invention of the hovercraft was hailed as an entirely new form of transport. But first he had to prove that it would work...

THE ORIGINAL EXPERIMENT

RATING: ✷✷ MODERATE

DANGER FROM SHARP TIN CANS

You will need
- 2 empty tin cans, one a little smaller than the other so that it fits inside the larger, leaving a gap
- Electric hairdryer – one that can be used on cool
- Gaffer tape
- Kitchen scales or more sensitive letter scales
- Double-sided tape or glue
- Strips of balsa or polystyrene ceiling tile, 5 x 25mm

What to do
1. Using tin-snips, cut a hole in the bottom of the larger can big enough for the hairdryer to fit tightly.
TAKE CARE: CUT TIN CAN IS EXTREMELY SHARP.
2. Insert the hairdryer and fix with gaffer tape to seal it on. Point the hairdryer/ can directly at the weighing platform of the kitchen scales. Set the dryer to cool, and turn on to maximum blow.
   What is the maximum thrust you can get with the tin at 1cm from the scales? Cockerell recorded 1lb (450g).
3. Now modify the set-up by suspending the smaller can inside the larger one. Using double-sided tape or glue (super glue would be good, but take care), attach strips of balsa or ceiling tile to the inside of the large can so that when the smaller can is inside, the rims of both cans are at the same level. Jam the smaller can into position.

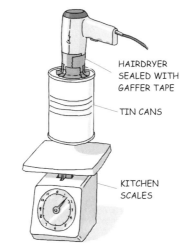

HAIRDRYER SEALED WITH GAFFER TAPE

TIN CANS

KITCHEN SCALES

4. Repeat the first experiment as above. How much thrust can you get this time? Sir Christopher got 3lb (1400g) – three times as much as for the can on its own.

> **WHY IT WORKS...** By forcing the air through a much narrower gap, its speed is greatly increased and so therefore is its momentum. Although the resistance is greater, this is outweighed by the greater thrust.

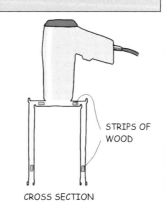

STRIPS OF WOOD

CROSS SECTION

BALLOON HOVERCRAFT

RATING: ✷✷ MODERATE

You will need
- Foil tray (e.g. for takeaway food)
- Polystyrene ceiling tile
- Used matchsticks
- Large balloon

What to do
1. Puncture a neat 5mm hole in the centre of the container, using a pen pushed from the outside.

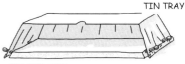

TIN TRAY

CEILING TILE

2. Make a 'baffle' from the ceiling tile, 2 or 3mm smaller all round than the opening in the container. Use matchsticks to simply wedge this in place, or stick the matchsticks into the container first for extra security. Try to get the gap between tile and container as even as possible.
3. To run the hovercraft, blow up the balloon. Holding the neck tightly, force the ring at the end into the hole – rather as you do in a balloon pump. Now you're ready to 'fly'!

NEWCOMEN ENGINE

In the early 18th century Thomas Newcomen made the first really useful steam engine, a vital spur to the Industrial Revolution.

DANGER FROM HEAT AND STEAM UNDER PRESSURE

Making your own engine is enormously satisfying. However, it is not for the faint of heart – it involves a fair amount of plumbing, metal work and soldering. Having said that, none of it is too difficult if you take it a step at a time.

You will need
- 1.5m length of 15mm-diameter copper pipe (for the piston/cylinder/pipework)

For the piston
- A foot pump 54mm in diameter (the kind for pumping up car tyres) – the internal piston is just the right size to fit a 54mm pipe
- Rubber O-ring (if necessary)
- Glue or a bent nail to fix shaft
- A bent tent peg

The 54mm copper piping is tricky to find – you could try a large plumbers' merchant, or possibly use the barrel of the foot pump instead.

For the cylinder
- 30cm length of 54mm-diameter copper pipe
- 2 small sheets of copper, about 50 x 50cm
- 2 x 54mm-diameter straight-through copper compression fittings
- Grease

These compression fittings are also tricky to find – however, they do make

life a lot easier. You should be able to buy copper sheeting at most craft and DIY shops.

For the pipework
- 1 x 15mm-diameter tee-pattern washing-machine tap (this has a straight-through part and a side arm with a valve on it)
- 1 x 15mm-diameter equal tee compression fitting
- 1 x 15mm-diameter straight-through washing-machine tap
- A pressure relief valve (about 3 bar) and appropriate fittings
- Depending on what fitting the valve has, you will need to find an adapter to allow you to attach it to a 15mm-diameter pipe.

Plus
- A good selection of tools, especially plumbing tools such as a bending spring, a blow torch, solder and flux; a circular cutter for cutting the 15mm pipe; a selection of files and drill bits and a decent hacksaw
- An assortment of bits of wood and planking to make the base
- An electric steam wallpaper stripper
- A plant water sprayer

We have tried various steam generators over the years, including modified pressure cookers, a home-made boiler and a camping kettle with pipe attached. These really could be dangerous – high-pressure steam is nasty stuff. Luckily, for this engine you only need steam at normal air pressure and the wallpaper stripper produces loads in complete safety.

What to do
Make the piston
1. This is the heart of the machine, the most important bit to get right. Start by dismantling the foot pump to extract the piston. Check that it fits into the 54mm pipe. If the fit is a bit loose you may need to fit another rubber O-ring on to the piston. It is vital that the piston fits really snugly or you will never be able to generate any force with your engine.

2. You should find that the piston is already fixed to a metal shaft. Unfortunately this shaft is usually too short. To extend it, fix the piston shaft into a 50cm length of the 15mm-diameter copper pipe. You can fix it in place with glue or, if the shaft has a hole drilled in it, a bent nail would do the trick. Drill a hole through the end of the copper piston shaft and attach the bent tent peg. This allows you to hook the engine up to whatever load takes your fancy, e.g. a bucket.

Make the cylinder
1. Cut two 54mm-diameter circles out of the two sheets of copper, and drill a 16mm hole in the dead centre of each. Place one of the plates on a heatproof surface, then solder a 2cm length of the 15mm copper pipe to the hole in the middle. Use plenty of flux, solder and a blowtorch to get a decent joint all the way around.

2. Now fit your two plates to the 54mm compression fittings. Remove the nuts from the fittings. To the top of each one solder one of your copper plates and screw the nut down on to the plate. Make sure the plate with the copper pipe soldered to it is fixed

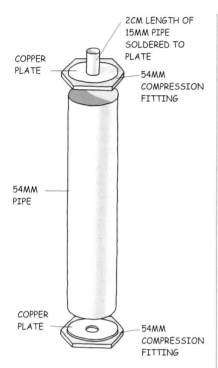

COPPER PLATE

2CM LENGTH OF 15MM PIPE SOLDERED TO PLATE

54MM COMPRESSION FITTING

54MM PIPE

COPPER PLATE

54MM COMPRESSION FITTING

so that the pipe is pointing out of the compression fitting.

3. Grease up the inside of the 54mm pipe, then assemble the cylinder by inserting the piston into the greased pipe. Use the compression fittings to seal the ends. Slide the copper piston shaft through the hole in the compression fitting without the short attached 15mm pipe. Fit the other compression fitting on to the other end of the 54mm pipe. Check that the piston slides easily in the cylinder – if not, you may need to re-grease the cylinder.

Make the pipework

1. To the short length of 15mm pipe poking out of the cylinder, attach the tee part of the tee-pattern washing-machine tap. To each of the other arms of the tap, attach a 10cm length of the 15mm diameter pipe. Tighten all the compression fittings.

2. Fit the straight-through washing-machine tap to the end of one of the 10cm pipes. To the end of the other 10cm-long pipe fit one end of the straight-though part of the equal tee compression fitting. Tighten the joint.

3. To the tee part (which should be pointing in the same direction as the cylinder) fit a 5cm length of 15mm copper pipe. To the end of this you will need to fit the safety valve using whatever connector it was supplied with.

Finishing off

Finally, connect up the wallpaper stripper. It is difficult to give precise details without knowing your model. You have to detach the stripping head from its flexible hose, and then find a way to connect the hose to a short length of 15mm pipe. This may be as easy as pushing the pipe into the hose. Fix the end of this pipe on to the last remaining hole in the equal tee.

Mounting

Your engine should be mounted on to a suitable base board. Try using a large sheet of thick plywood with suitable lengths of fence post used as supports for the piston. The cylinder itself is best mounted vertically. If you mount the cylinder so that the piston pulls vertically up, you can easily measure the force produced by the engine by simply hooking a bucket on to the end of the piston and the vertical pull will re-set the piston.

Steaming up and making a power stroke

1. First fill the stripper with water, and switch on. Push the piston into the cylinder as far as it will go, and close the valve leading into the cylinder (if your piston is a really good fit it should stay there).

2. Open the other tap – this is your pressure release valve and must be open while you start steaming up, or the safety valve will blow open. Once steam is freely pouring out of the end valve, open up the cylinder valve and pull the piston down to fill the chamber with steam. Push all the steam out of the cylinder and repeat this process a few times, until the cylinder is really hot. With the cylinder pushed in, close off the end valve, leaving the cylinder valve open. The piston should slowly be pulled down, as it fills with steam. Once the piston is fully out, shut the cylinder valve and open the outlet valve. The cylinder is now full of steam.

3. To generate a power stroke, simply spray the outside of the steam-filled cylinder with cold water.

DID YOU KNOW?

James Watt is often credited with inventing the steam engine, although this is in fact nonsense. By the time he was born in 1736 there were already dozens of huge, lumbering Newcomen engines pumping water out of coal mines – the first had been installed in 1712. Watt realized that alternately heating and cooling the huge metal cylinder was a terrible waste of heat, and designed a separate condenser so that the cylinder was always hot. He finally got his engine to work in 1775, and it

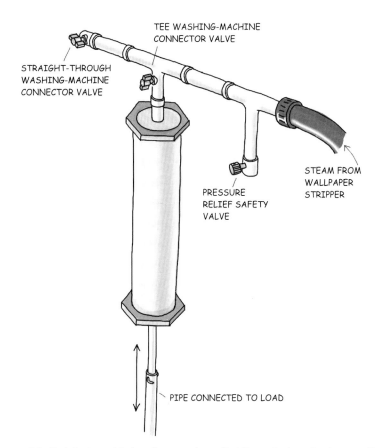

TEE WASHING-MACHINE
CONNECTOR VALVE

STRAIGHT-THROUGH
WASHING-MACHINE
CONNECTOR VALVE

STEAM FROM
WALLPAPER
STRIPPER

PRESSURE
RELIEF SAFETY
VALVE

PIPE CONNECTED TO LOAD

the condenser. The steam will condense – but without cooling the cylinder.

THOMAS NEWCOMEN

Thomas Newcomen came from Dartmouth in Devon and although he was ingenious he wasn't much of a businessman and lacked the influence to exploit his great invention. Largely to blame was another Devon inventor, Thomas Savery, who had invented a similar device at about the same time. Savery's 'Miner's Friend' was not much good – it partly used high-pressure steam and his solder joints weren't really up to the job. The job in question was pumping water out of tin mines: tin was so valuable that it was worth paying to keep flooded mines open. Newcomen, whose engine did work, should have made a killing. The trouble was that while Savery had a hopeless engine, he had a brilliant patent. The King (a personal friend) had granted Savery a patent for 'engines for raising water by the impellant force of fire' – which meant any engine that used fire to raise water, including Newcomen's. So poor old Thomas Newcomen was forced to go into partnership with Savery, and never made the fortune he deserved.

was this that first provided real power for factories.

BUILD A WATT ENGINE

You could modify your Newcomen engine just as James Watt did. Between the cylinder and the straight-through valve add another tee connector joined to a long pipe submerged in cold water – your condenser. We made a zig-zag pipe to save space. When you want to make a power stroke, instead of cooling the cylinder, simply open the new tee and draw the little steam (we used a syringe on the end of the zig-zag tube) from the cylinder into

WHY IT WORKS... As the cylinder cools, it cools the steam inside. The steam almost instantly turns into water – which takes up only one-thousandth of the space and so generates a vacuum. Atmospheric pressure then pushes the piston into the cylinder, creating the power stroke. Your piston should be able to move at least 3.5kg in weight. Once your wallpaper stripper is boiling away furiously, it should be possible to produce repeated power strokes without too much effort.

FARADAY'S CANDLES

Michael Faraday introduced the Christmas Lectures at the Royal Institution, and in 1855 his topic was the chemical history of a candle. Here are a couple of his lovely demonstrations.

RELIGHTING A CANDLE

RATING: ✳ EASY!

DANGER FROM NAKED FLAME

You will need

- One candle or preferably two. Ordinary household candles are fine as long as you can keep them upright; short chunky ones are better, but nightlights have thin wicks and so only small flames
- For relighting the candles, preferably a wax taper (or matches/lighter)
- A candle snuffer (see below for how to make your own)

It's a good idea to have a second candle and keep it alight. You can relight everything from this, so you don't need to use matches or a lighter so much.

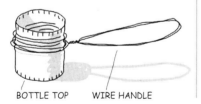

BOTTLE TOP   WIRE HANDLE

To make a candle snuffer, fix a wire handle to a metal screwtop from an old bottle, as shown in the diagram. (You can blow the flame out, but this makes the relighting much more difficult because the air gets turbulent.)

For a really smart way to blow a candle out, see the Vortex gun on page 18. However, you have to be dextrous to blow out and relight a flame like this.

What to do

1. Stand the candles a few centimetres apart and light them both.
2. Light the wax taper if you have one. If not, use a lighter or a match – light it from one of the candles.

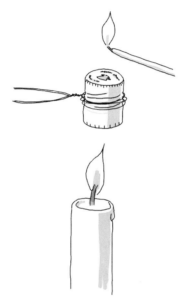

3. Snuff out one candle by gently lowering the bottle-top over the flame.

As soon as the flame goes out you will see white stuff rising from the wick. Using your taper or lighter, you can light this a couple of centimetres or so above the wick and see the flame jump back down the stream to the wick.

For best results, use a candle with a fat wick to give a big flame. Let it burn for a few minutes before putting it out.

Hold your burning taper a couple of centimetres or more above the wick and carefully lower your snuffer. As soon as the flame is out move the snuffer away sideways, keeping the taper in position.

More things to try

RATING: ✳✳ MODERATE

DANGER FROM NAKED FLAME

● To make a bigger flame: find a piece of string – the fatter the better. (This has to be the absorbent kind, like white kitchen string, not nylon or plastic.) Drill or poke a little hole beside the existing wick, and push the other piece of string in. Or you could

---

WHY IT WORKS... Candles are made of wax. The best ones are made of beeswax, which burns with a fine even flame. Solid wax will not burn – try holding the bottom of one candle over the flame of another. Liquid wax will not burn either; in fact, too much of it will put the flame out. The only sort of wax that burns is wax vapour.

When a candle burns, the heat from the flame melts the solid wax below, and makes a pool of liquid wax. This soaks into the wick, and is pulled upwards by capillary action (there's more about this action on pages 48–9 and 55). When the wax gets into the flame, it is heated so much that it vaporizes. Wax vapour, or wax gas, rises from the wick and burns in the air. The wick is only a piece of string, but it acts as an engine for getting the wax into the vapour phase.

try two wicks side by side. Either way, you should get a bigger flame.

Can you get the flame to jump back from higher up, when the flame is bigger?

● What about making other kinds of candle? Almost any fat or oil will burn as long as it has a wick to get it into the vapour phase. Try making candles using any of these: butter, margarine, lard, cooking oil, mineral oil (e.g. bike oil) or soap. Can you think of anything else you could use?

WARNING!
DO NOT TRY PETROL, LIGHTER FLUID OR NAIL-VARNISH REMOVER. THESE ARE ALL HIGHLY FLAMMABLE AND WOULD BE VERY DANGEROUS!

For the wick you will need something like kitchen string, as described above.

Take a little piece of butter on a plate or spoon. Poke a hole in it with a matchstick or a fork. Cut a couple of centimetres of string, and push half of it into the hole, so that the free end stands up.

● To make a 'liquid candle', or oil lamp, take a piece of string approximately 5 centimetres long and tie a knot in it so that it will lie with one end in the air. Put this in a milk-bottle top, or a spoon, and pour on half a teaspoonful of oil.

Now try lighting the wick. Will the stuff burn without the wick?

Fuels that vaporize easily, such as petrol, will burn as liquids, but oils need wicks to lift them into the vapour state.

TWO FLAMES FROM ONE CANDLE

RATING: ✳✳✳ CHALLENGING

DANGER FROM NAKED FLAME AND HOT TUBE

You will need
● A glass tube about 4mm or 5mm in diameter and 6–10cm long
● A piece of wire to make a holder for the tube
● Two candles
● Taper, matches or a lighter

What to do
1. Wrap one end of the wire tightly around the glass tube and the other around one candle, and fix the glass tube so that its bottom end is about 3–5mm above the top of the wick, and the tube is angled up at about 45 degrees.
2. Light the candle. The bottom end of the glass tube should be in the middle of the dark part of the flame.
3. Place the second candle under the glass tube and light it, so that the flame heats the tube.
4. Hold a lighted taper, lighter or match to the top of the glass tube. You should be able to get another flame here.

The bottom of the tube needs to be close to the top of the wick, well down in the flame. The glass tube needs to be hot.

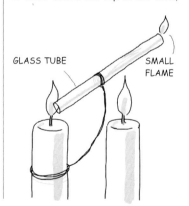

GLASS TUBE

SMALL FLAME

DID YOU KNOW?
Michael Faraday never went to school, and yet he became one of the greatest scientists of his time.

WHY IT WORKS... The heat from the first candle turns the wax into liquid and then into vapour, which streams upward from the top of the wick. When the glass tube is in the right place, some of this vapour will flow up the tube. If the tube is cold, the vapour will condense inside, and form liquid and solid wax, but if the tube is hot the vapour will flow up to the top, and burn there when you light it. Then you have two flames from one candle.

VORTEX GUN & SMOKE-RING CANNON

The 19th-century scientist Charles Vernon Boys came up with this idea. Legend has it that he used a smoke-ring cannon to shoot stink-bomb gas at passers-by on London's Victoria Street. The vortex gun also makes a fine tool for blowing out candles.

VORTEX GUN

RATING: ✳ EASY!

DANGER FROM NAKED FLAME

You will need
- A long tube of crisps (eat the crisps)
- A sharp craft knife
- A two-pence piece
- A large balloon
- Strong tape (gaffer tape or electrician's tape)
- A candle on a candlestick

What to do
1. To start, you need to cut the metal base off the tube. The simplest way to do this is to cut through the cardboard, just above the metal bit, with a sharp craft knife. Cunning people could try using a tin opener to cut a hole in the metal bit of the

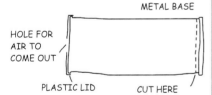

HOLE FOR AIR TO COME OUT

METAL BASE

PLASTIC LID   CUT HERE

base, but it's not easy. Take care with the knife – it's easy to slip and hurt yourself.
2. You also need to cut a circular hole in the centre of the plastic lid. It is important that the hole is centrally placed, not too big, and is a smooth circle. You won't get the right effect with an off-centre jagged hole. The simplest way to do this is to use a two-pence piece to mark the hole, then either cut the plastic with a

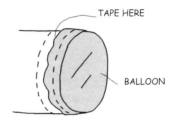

TAPE HERE

BALLOON

craft knife or with a small pair of scissors.
3. Now cut about 2cm from the neck of the large balloon and stretch the remaining bit of balloon over the open bottom of the tube until it is fairly taut. This is a bit tricky and may require a second person to hold the tube. Wrap a bit of tape around the tube to hold the balloon on (it has a tendency to ping off otherwise).

Firing the gun
You can see what the gun can do by trying to blow out a candle with it. Hold the tube with the plastic lid about 15cm from the flame. By firmly tapping the rubber skin at the other end, air will shoot out through the hole in a way that might surprise you – it takes longer than you expect to reach its target.

Unless you get lucky, you will most likely just see the flame flicker. Keep trying by carefully changing the angle of the gun. Assuming the room you are in has no draughts, the 'packet' of air will go in a straight line. You should be aiming at the wick of the candle. Once you have blown the candle out, move back a little and try again. Can you beat my 2-metre record?

More things to try
● You can show what's going on by using a row of candles very close to each other. Aim your vortex gun, and see which ones go out.
● You could do the same thing in two dimensions. Arrange an array of candles at different heights – say, four rows of four. See if you can blow out a ring of candles.

BE WARNED: THIS CAN GET VERY HOT AND YOU MUST ENSURE THE CANDLES ARE FIRMLY FIXED.

When we tried it the candles burned very fiercely, the higher ones fanned by the up-draught from the candles below!

WHY IT WORKS... When air is forced through the circular hole, the air at the edge is held back by friction and therefore moves more slowly. By contrast, the air in the middle of the hole is shooting forward with increased speed. This sets up a rotation from centre to edge (you can see this happening with the smoke-rings in the next demonstration). The motion keeps going because the increased speed in the centre results in reduced pressure, which tends to suck in the higher-pressure air from the edges. The ring keeps going until the energy you gave it by pushing the balloon skin runs out.

SMOKE-RING CANNON

RATING: ✳✳ MODERATE

You will need
- A vortex gun like the one opposite
- A smoke generator

How do you fill the cannon with smoke? Smoke pellets are available from most fancy-dress and joke shops – but they make much more smoke than you need. Really to do justice to a smoke pellet you should build the giant cannon suggested below.

You could try using an old tin can, packed full of smouldering cardboard and a funnel to get the smoke into the cannon – but it is hard to control the flames. Unfortunately the most convenient way of making smoke is to use a cigarette, something I normally would not condone. In the side of the cannon make a small hole that just fits a cigarette snugly. Insert a lit cigarette, careful not to set fire to the cardboard tube, and you have a long-lasting, slow production of smoke.

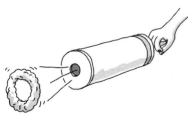

What to do
1. To make a smoke-ring, all you need do is fill the cannon with smoke and tap on the base. The smoke-rings can be made to go faster by tapping harder on the rubber skin.

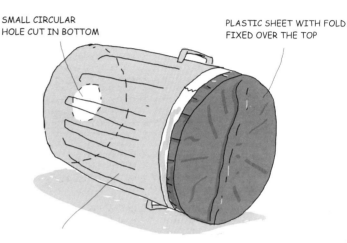

SMALL CIRCULAR HOLE CUT IN BOTTOM

PLASTIC SHEET WITH FOLD FIXED OVER THE TOP

PLASTIC BARREL, BUCKET, DUSTBIN ETC.

Bigger and better
However, you don't need to stop there. In the programme, I used an old plastic beer barrel with a bit of rubber pond lining glued to the base to make really big smoke-rings. Other suitable containers might include plastic wastepaper bins or dustbins. (One viewer even claims to have done this with a garden shed – though that seems a bit over the top.)

If you try the bigger versions, you will need to attach the rubber in such a way that it is a bit baggy. This is to allow it to pump smoke out of the barrel, and also to give you something to get hold of. I suggest making a fold in the sheet. You might get away with a bin-liner, though the stiffer sheet seemed to work rather well. Fix the sheet by gluing or taping. For this size of cannon, a smoke pellet is ideal.

I managed to shoot really good smoke-rings across the auditorium of the Royal Institution. Given still air, you ought to get them to travel over 10 metres.

DID YOU KNOW?
Charles Vernon Boys was born in Rutland in 1855, and became a brilliant scientist, ranging over a vast variety of different areas. As well as making the most accurate gas-meter in the world, he was also the king of bubbles, as you can read on page 29, with bubble tricks for you to try.

Sir Richard Paget summed him up like this. Why was it, Paget asked, that Boys would tackle:

'A problem older men would shirk
Yet solve the task, and make it work
By means that no one else employs?
The answer is: Boys will be Boys.'

ENGINES IN A CAN

Here is a chance to demonstrate dramatically the power of the atmosphere and the power of steam – the driving forces of man's earliest engines. One was invented by Hero of Alexandria (rather a good name for an inventor in this book) around 100 BC.

ATMOSPHERIC ENGINE

RATING: ✱✱ MODERATE

DANGER FROM HOT WATER AND FLAME

You will need

- Washing-up bowl filled nearly to the brim with cold water
- Empty fizzy drink can – the very thin aluminium kind
- Tongs to grip the can firmly
- Oven gloves
- Heat source to boil water in the can. You might get away with a candle, but we used a camping gas stove. Most of you will use the stove in your kitchen – but take great care.

What to do

1. Place the washing-up bowl of water close to the stove – you will need to get the can from stove to bowl in a second or two. Put a little water into the can through the hole in the top – about 5mm in the bottom of the can should be enough.

2. Grip the can with the tongs, and have an oven glove on the other hand in case the can slips and you need to grab it. Heat the can over the flame until the water boils – it makes sense to use the lowest flame possible.

MAKE SURE THERE IS SOME WATER IN THE CAN OR YOU WILL MELT IT AND MAKE A MESS ON THE COOKER.

3. When the water is boiling merrily, the can is filled with steam. When it comes out of the hole, it's time to make your move. Using the tongs,

move the still-boiling can over the washing-up bowl, turn it over so that the hole is at the bottom, and plunge it into the water until the end of the can with the hole in is fully immersed.

4. Bam! The can should almost instantly collapse in on itself as if you had stamped on it. You may have to practise the flick of the wrist needed to make the demonstration work properly.

If the can doesn't collapse, either there wasn't enough steam in it in the first place, or you let it cool too much before plunging it into the water.

DID YOU KNOW?

In Britain, Thomas Savery of Shilston Barton in Devon was the first to make an engine based on this principle, inspired by watching the wine boil in the bottom of a bottle he had tossed into the fire. His Miner's Friend, or 'Engine for raising water by the impellant force of fire' as his patent of 1698 called it, was used to pump water out of mines. It was never very good, but pioneered the use of atmospheric engines that helped to launch the Industrial Revolution. Like the tin can, it worked by boiling water inside a sealed container and then condensing the steam suddenly.

HERO OF ALEXANDRIA FIZZY-DRINK CAN ENGINE

RATING: ✱✱ MODERATE

DANGER FROM HOT WATER, STEAM AND FLAME

Hero (or Heron) of Alexandria is supposed to have made an engine based on this principle in about 100 BC. It is really a jet engine, and you need to take care with the jets of hot steam. It can also be rather messy.

You will need

- UNOPENED can of fizzy drink with ring-pull tab intact
- Bucket
- Sharp tool to make a hole – the spike of an old pair of school compasses is ideal
- Thin string, about 50cm long
- Piece of wood to act as support – a wooden spoon would do
- Washing-up bowl of water
- Gas flame to heat can, from domestic or camping stove
- Oven gloves

What to do

1. This is the messy bit. Hold the can inside the bucket and puncture it with your spike about 1cm below the rim of the can. Push the spike in, and

WHY IT WORKS... When the water boils the can fills with steam, pushing out all the air. When the hot steam hits the cold water, it immediately condenses and becomes water again. This creates a partial vacuum inside the can. But the air, which was previously pressing on the can from inside and out, now presses only from the outside and squashes the can flat. The trick depends upon the fairly small hole in a normal drink can, which stops water from the bowl filling the can quickly enough to replace the steam.

then turn it to the right 90 degrees until it lies flat against the side of the can. This slightly distorts the can and creates a jet facing to the right. As you do this, of course, drink will come spurting out into the bucket.

2. THEN PUSH FLAT AGAINST SIDE OF CAN

1. PUSH SPIKE STRAIGHT INTO CAN TO MAKE HOLE

CLOSE UP VIEW OF JET

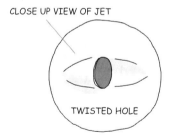

TWISTED HOLE

2. Now turn the can round and make another hole on the opposite side, also 1cm from the top, twisting the spike to the right to make another sideways-facing hole.

3. Now you have to get rid of the drink. Holding the can upside-down in the bucket, give it a good shake. All the drink will be powered out of the holes in the can by the carbon dioxide.

4. When the can is empty, tie a piece of string round the ring-pull, which must be intact. Slide the string as close towards the centre as you are able, in order to suspend the can as upright as possible. This is easier with some cans than others. Tie the other end of the string to the wooden support.

5. Now dip the can into the bowl of water, submerging one hole, until you get about 2cm depth of water in the bottom. This is a bit difficult to judge, of course, but doesn't have to be accurate.

6. Light the gas and put on the oven gloves. Holding the wooden support, dangle the can over the flame. Heat until the water boils. When steam comes out of the jet-holes, it will spin the can, making this a model of the first jet engine we know about.

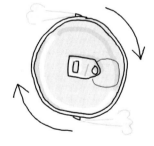

HEAT

NEVER HEAT A SEALED CAN – ONLY ONE WITH HOLES IN IT.

You may have difficulty running the engine out of doors. It is surprising how a breeze can blow the flame and cool the can. Even in the baking heat of Alexandria in Egypt, an ordinary small camping stove was no good, and we had to buy a rather violent local model. One viewer reports trying this over a barbecue, which sounds good – but take care, they are very hot.

DID YOU KNOW?

At normal air pressure, when water turns to steam it increases in volume by 1000 times.

WHY IT WORKS... Isaac Newton found that for every action there is an equal action in the opposite direction: this is his Third Law of Motion. So throwing molecules of steam out of the can in one direction results in the can moving in the opposite direction. (There is a different demonstration of this law on page 24.) In space, astronauts are able to use this principle to move themselves about easily just by using a burst of gas. Back on earth they need the near-explosive power of rocket fuel to get a spacecraft into orbit.

CHAPTER TWO
AIR POWER

JETS

In 1928 Frank Whittle wrote an essay on 'Future Developments in Aircraft Design'. He argued that to fly fast and far, you must fly so high that the air is thin and air resistance is minimal, and to do that you would need a jet engine. Whittle had his first jet engine running on 12 April 1937, and got the first jet aircraft into the air on 15 May 1941. Here's a DIY version of a simple jet engine.

A BALLOON-POWERED BOAT

RATING: ☀ EASY!

You will need
- A balloon
- A small boat (see below)
- Tape
- Bath, basin or pond

You can buy cheap little plastic boats from toyshops, and one of those is ideal. Or you can make your own, from plastic, wood, metal or even paper. All boats work better if they are fairly long and narrow, and preferably pointed at the bows (front).

Or you could make a plastic boat from an old margarine tub, or a bottle that used to contain fizzy drink, washing-up liquid or shampoo – or even an empty toothpaste tube. As long as it floats, it will work. If it rolls upside-down, weight it down with something like plasticine, or with a nail taped to the bottom.

What to do
1. Tape the balloon to the top of the boat so that the nozzle points back over the stern and there is still space for you to blow it up.
2. Blow the balloon up, and then holding the nozzle shut, put the boat in the water and let go. As the balloon deflates, the boat should sail across the water.

Don't expect it to go very fast – or very far. However, the lighter the boat, the better it will go.

More things to try
● Does the boat go faster – or further – if you inflate the balloon more?
● What about using a bigger balloon?
● Does the shape of the balloon make any difference?
● What about fitting the balloon to a tube, in order to make the air come out in a straight line? You could use the tube from an old felt-tip or marker pen, or use about 10cm of hosepipe. Does the boat go better?
● Can you find or make a model car that runs smoothly enough to be powered by balloon?

CONGREVE ROCKETS

William Congreve (born in 1772) developed the military rocket – a simple form of jet engine – and in 1806 he had a chance to attack the French navy at Boulogne, using 14-kilogram rockets with about 5-metre guide sticks.

Unfortunately, all 200 rockets were blown off course by the wind. They missed every boat in the harbour and set fire to the town, causing a lot of damage. On the battlefield the rockets scared the horses on both sides, but in spite of these disasters, Wellington did reluctantly take a rocket brigade to Waterloo in 1815.

William Congreve is also remembered for the Congreve Clock, an ingenious device which instead of a pendulum as regulator, used a ball-bearing rolling in a zig-zag groove down an inclined plane, which tips first one way and then the other.

WHY IT WORKS... Why does the boat go forward? This is another demonstration of Newton's Third Law of Motion (also shown on page 21), which says that for every action there is an equal and opposite reaction. When you pick up something heavy, your muscles are raising the weight and at the same time your feet are pressing on the floor with an extra force equal to that weight.

The elastic balloon squeezes the air out of the nozzle and, as it gives the air a push backwards, the rubber is pushed forwards, and takes the boat with it. This is the principle of the rocket engine. The air rushing out of the nozzle does not have to push against the air behind. This sort of engine works in the vacuum of space, where there is nothing to push against, which is why spacecraft have rocket engines.

AIR-POWERED RAILWAYS

Pneumatic and atmospheric railways were 19th-century inventions that carried goods and passengers by using differences in air pressure. The pneumatic system is easier to make – you might like to tackle the simple tabletop model. Or, if you're feeling more ambitious, try making an atmospheric railway that's big enough to carry a person.

PNEUMATIC RAILWAY

RATING: ✳ EASY!

You will need
- A stiff plastic pipe about a metre long
- A piston that will just fit inside the tube, to act as a model 'carriage'
- A vacuum cleaner fitted with a flexible tube

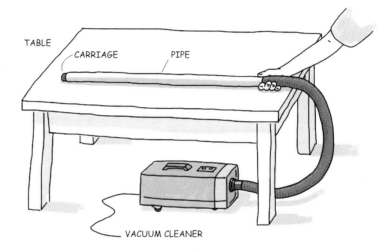

The most difficult thing is finding a piston that just fits. I was using a plastic pipe intended for plumbing, and after much trial and error found a plastic drum of goldfish food that slid comfortably down the pipe without any visible gap round the edge.

What to do
1. Lay the pipe on a table, place the 'carriage' in one end, switch on the vacuum cleaner, and apply the nozzle of the flexible tube to the end of the pipe. The carriage will rocket down the tube towards the vacuum cleaner.
2. You may have to adapt the end of the nozzle to fit your pipe, but usually you can do this by using your fingers to plug any major gaps, or by using clingfilm wrapped around the join.

MODEL ATMOSPHERIC RAILWAY

To turn this into a model atmospheric railway, find a small model railway carriage (or anything else with wheels) and tie it a metre behind the piston with a long piece of string.

More things to try
● Find out which of two vacuum cleaners is the more powerful. Fit both cleaners with their flexible hoses, without any tools on the end. Switch them on. Put a ping-pong ball on the end of one so that atmospheric pressure pushes it hard against the nozzle. Now see whether you can pull the ball off using the other vacuum cleaner. The more powerful machine will always win this tug of war. Don't do this for too long, though, or the motor will overheat.
● Measure the vacuum in your vacuum cleaner. Put on your kitchen scales a large tin of beans, or some other object with a mass of about 1kg and a smooth flat top. Note the weight. Now try to lift the can using the nozzle on the end of the vacuum cleaner hose. You probably won't be able to lift a large can of beans, but you will be able to reduce its apparent weight. Atmospheric pressure is now helping to lift the can because there is low pressure in the vacuum cleaner nozzle. See how much you can reduce the weight of the can.

Now calculate the area of the circular nozzle of the vacuum cleaner. My ageing vacuum cleaner will reduce the weight of the beans by 900 grams, which is equivalent to a force of 9 newtons. The area of the nozzle works out at 6 sq cm. Therefore the reduction in pressure inside is given by 9/6 or 1.5 newtons per sq cm.

This pressure reduction is only about 15 per cent of atmospheric pressure, so the vacuum cleaner doesn't make a good vacuum. The fan inside is designed to move a lot of air – we should really call them 'air-flow cleaners' instead of vacuum cleaners!

DID YOU KNOW?

In New York, Alfred Ely Beach, former owner of the *New York Sun* and *Scientific American*, built a model pneumatic railway for the American Institute Fair in 1867, and a real one, which ran for 100 yards under Broadway.

This opened on 28 February 1870 with tremendous celebrations. The *New York Times* reported:

'Certainly the most novel, if not the most successful, enterprise that New York has seen for many a day is the pneumatic tunnel under Broadway.'

The whole thing was certainly grand. The waiting room was decorated with chandeliers, a fountain, a grand piano, and a grandfather clock. Unfortunately, Beach was unable to extend the line for political and economic reasons.

However, pneumatic despatch became quite common in the second half of the nineteenth century. The Post Office despatched telegrams all over London along pneumatic tubes about 5cm in diameter, using positive air pressure to despatch them out from the centre and negative air pressure from a vacuum pump to bring them back.

In many department stores pneumatic tubes carried cash from customers to the cash desk, and came back with the change.

ATMOSPHERIC RAILWAY TO RIDE ON

RATING: ✳✳✳✳ DIFFICULT

You will need
- One or two vacuum cleaners
- Pipe and piston, as on page 25
- Model 'carriage', with free-running wheels and a suspension high enough to run above the pipe

See below for how we made our versions of the pipe/piston and carriage.

The vacuum cleaner(s) will cause the piston to travel along the pipe, as with the Pneumatic railway. However, for the Atmospheric railway you need a slot along the top of the pipe. Through this slot the piston is connected to the carriage, which runs along above the pipe. The real difficulty is finding rigid pipes. We tried making small versions using plastic drainpipe, but when you apply a vacuum cleaner this tends to curl up inside itself and refuses to let the piston pass along.

What to do
1. For our pipe, we used cardboard tubes 15cm in diameter and 2m long. Tape six of these tubes end to end, passing strong brown parcel tape round the circumference to make an airtight seal at each join.
2. With a jig-saw cut a slot along the whole length, about 2mm wide. Close the far end of the pipe with a lid, and drill 3cm holes to fit the nozzles of the two vacuum cleaners. To maintain the vacuum ahead of the piston press a line of those removable sticky notes along one side of the slot.

WHY IT WORKS... When you switch on the cleaner you create a part vacuum inside the machine and in the flexible tube. You can feel the vacuum if you put your hand over the end of the nozzle. This is not a complete vacuum; the pump in a vacuum cleaner is designed to give a rapid flow of air in order to shift dust from the floor, rather than to remove all the air inside. Even running flat out with the end of the nozzle blocked, it will probably remove only a quarter of the air from inside, but this is plenty to drive the pneumatic railway.

This is how the difference in air pressure takes effect. Atmospheric pressure is about 10 newtons per square centimetre (or 15 pounds per square inch). One newton is about the weight of an apple; so the atmosphere pushes down on each square centimetre of every surface with a force about equal to the weight of ten apples.

The lid of my drum of goldfish food has a diameter of 4.5cm, which means an area of about 16 sq cm. Therefore the force of the atmosphere on the top of the lid is about 160 newtons. Normally this would be balanced by an equal force on the bottom of the drum, but if the vacuum cleaner removes a quarter of the air from the pipe, then there will be a net force of 40 newtons (one-quarter of 160) pushing the piston along the cylinder. This is a much bigger force than the weight of the piston; so the piston moves quickly. (And the lighter the piston, the faster it moves.)

Some pneumatic systems use positive air pressure from an air compressor to push the piston, but the principle remains the same: the piston is driven along the cylinder because the air pressure is higher behind than in front.

CARRIAGE   TOW BAR
           OR STRING   SLOT

PISTON

VACUUM CLEANER

3. The piston was made from an old paint can which just fitted in the tubes. Pass a long bolt through the middle of it, and attach a string (thin enough to pass smoothly through the slot) to the bolt. For a test run without a driver, tie the string to the carriage as shown. The driver – or 'passenger' – holds this string (prepared to let go of it in a hurry if necessary!).

4. The carriage was made from four bicycle wheels and a minimal structure of four pieces of wood. The carriage must be high enough off the ground to run over the pipe, but you might be able to use an old pram or

pushchair, or build up from a pair of skateboards.

5. With the passenger sitting on the carriage, holding the string tied to the piston, attach the vacuum cleaner nozzles, switch on the motors – and bon voyage!

More things to try

● Although the plastic pipe we tried

wasn't much good, I have heard of successful atmospheric railways made using plastic drain pipe or waste pipe (the 1½-inch or 37mm sort). You'll just have to experiment.

● We needed a huge pipe and two vacuum cleaners because we had to pull a vast weight (guess who) over rather rough ground. If you are running indoors on a smooth floor you may well get away with one vac and the smaller pipe.

DID YOU KNOW?

Isambard Kingdom Brunel built 20 miles of atmospheric railway down the coast from Exeter to Newton Abbot, and it ran as a popular passenger service in 1847 and 1848. Unfortunately, it was then closed down after a series of technical and financial problems.

WHY IT WORKS... The principle is exactly the same as for the pneumatic railway. The vacuum cleaners remove part of the air from the pipe, creating a low pressure inside, and the pressure of the atmosphere pushes the piston along the tube, pulling the carriage behind it. So it really is an atmospheric railway, since it is driven by atmospheric pressure.

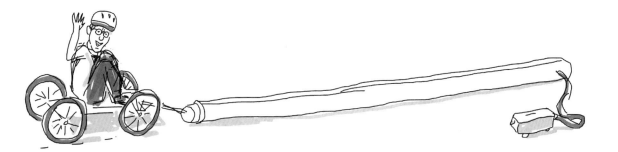

BUBBLE POWER

Bubbles look pretty and are easy to make, but they also provide excellent demonstrations of surface tension, the size of molecules, and currents in the air.

BUBBLE HOLE

RATING: ✳ EASY!

You will need
- Bubble solution (see below)
- A flat tray with a rim, or a roasting tin
- 40–50cm of stiff wire – an old coat-hanger is ideal
- 40–50cm of cotton thread
- A couple of long tools, e.g. pencils, pens, nails, chopsticks

You can buy bubble solution, but you might prefer to make your own. Buy a small bottle of glycérine (or glycerol, which is the same thing) from a chemist. Into a closable container (e.g. an old jam jar) put half a teaspoon of glycerine, five teaspoons of washing-up liquid and 300ml of water. This needs to mix really well, so whisk or shake it thoroughly – but don't get too much air in or it will turn to froth. Then leave it to stand over-night, and it will make better bubbles.

CAUTION!
BUBBLE MIXTURE IS MESSY STUFF. YOU ARE STRONGLY ADVISED TO WORK OUTSIDE, OR IN THE KITCHEN SINK OR THE BATH, IN ORDER TO CONTAIN THE MESS!

What to do
1. Put enough bubble solution into the tray to form a shallow pool. Bend the wire to make a flat, closed loop with a stand. Tie 20cm of the cotton in a loop, and attach this, using more cotton, to the top and bottom of the wire loop.
2. Dip the wire loop and cotton in the bubble mixture so that the whole lot,

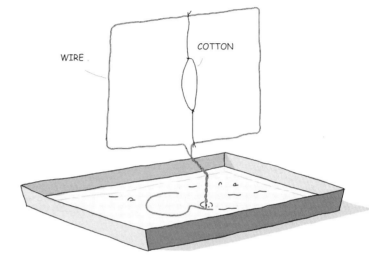

WIRE   COTTON

apart from the stand, is wet, and set it upright. You should have a soap film right across the wire loop with the cotton hanging in it. If the soap film keeps breaking, bind all round the wire with wool or kitchen string.
3. If the cotton loop is closed up, wet your pencils (or equivalent) in the bubble solution, and use them to tease it open. As long as they are wet they will not break the film. Then touch the film inside the loop with a dry finger. Instantly the film will break and the cotton loop will make a perfect circle.

More things to try
● Wet a finger in the soap solution (or use a wetted pencil), put it through the hole you have made, and push the side of the circle sideways or down – along the direction of the film. What do you see and feel? How can you explain the force holding back your finger or pencil?
● What will happen if you make a loop within a loop, or two loops together, and pop the film first in one and then in the other? Make predictions, then try it.

WHY IT WORKS... Soap films such as this one have surface tension, a force acting along the surface to make it behave as though it is stretched. As a result of this force, the surface of any soap film will always take up the smallest possible area. Bubbles are spheres because a sphere gives the surface the smallest possible area for the amount of air inside; any other shape would have more surface. When you break the film inside the cotton loop the rest of the film wants to make its area as small as possible; so it pulls the cotton loop out in all directions. The surface tension is the same in all directions, so the loop is pulled equally all round to make a perfect circle.

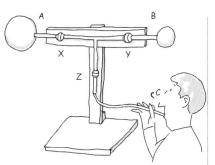

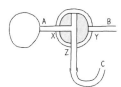

OPEN X AND Z AND SHUT Y. DIP A INTO SOLUTION AND BLOW INTO C UNTIL YOU HAVE A BUBBLE 6-8 CM

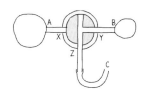

SHUT X OPEN Y. DIP B INTO SOLUTION AND BLOW INTO C UNTIL BUBBLE AT THE END OF B IS 4-6 CM

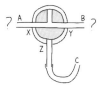

CLOSE Y AND Z. OPEN X AND Y. WHAT HAPPENS?

BUBBLE TRANSFER

You will need

- Flexible plastic or rubber tubing, about 5mm diameter
- Stopcocks or other means of closing the tubes
- A stand to hold it off the ground – or you could tape the tubes to the tops of food cans
- Bubble solution (either bought or home-made – see opposite page)

The best gadget for closing the tubes is the three-way stopcock, above right. Or you could use three normal stop-cocks, joining them with a T-shaped piece of flexible tubing, as shown.

Alternatively, you could close each tube by doubling it over and clamping it with a bulldog clip or a clothes peg.

What to do

1. Open the tube at x and z, and shut y. Dip end A in the bubble solution, and blow gently into C until you have a bubble about 6–8cm in diameter on the end of the tube. Shut x. Open y. Carefully dip end B into the bubble solution (or smear a little solution over it), and blow gently until on the end of B you have a bubble 4–6cm

in diameter (i.e. smaller than on A). Close y. Close z.

2. What will happen when you open x and y? Will the big bubble blow up the small bubble? Will they both become the same size? Make a prediction then open x and y and see what happens.

More things to try

● Try doing the same experiment but using balloons instead of bubbles. Do you think the results will be the same?
● Try blowing a bubble inside a bubble. Use the apparatus above, but first insert a longer piece of tubing (say 25cm) between the T-piece and stopcock B. Blow a bubble of 5 or 6cm diameter on the end of A, and close x. Wet end B thoroughly in bubble solution, and push it carefully

through the side of the first bubble; it should go in without puncturing the bubble, as long as it is wet. Then open y and gently blow a bubble on end B, inside the first bubble.
● What happens to the size of the first bubble? What would you expect to happen if you close z and open x to connect the two bubbles?

CHARLES VERNON BOYS

The brilliantly versatile scientist Charles Vernon Boys was king of the bubbles, too. When not engaged in grander projects, he'd investigate the science of bubbles – and in 1902 he wrote one of the best books about them: *Bubbles and the Forces That Mould Them.* (We also meet him on pages 18 and 19.)

WHY IT WORKS... When you blow up a bubble you increase the air pressure inside. The surface tension tries to pull the bubble smaller, and the extra air pressure inside pushes out just enough to keep it inflated.

To make a bubble bigger you have to put in more air, but the pressure actually drops! The air pressure is greater inside smaller bubbles than in bigger ones. You may have noticed that blowing up a balloon is most difficult at the beginning, when the balloon is smallest, and gets easier as the balloon grows, until the rubber is stretched as far as it will comfortably go. The air pressure is highest inside the smallest balloons, as it is inside the smallest bubbles.

As the pressure in the smaller bubble is higher, when you connect the two bubbles the small one will always blow up the bigger one.

CATCH THE WIND

Here are some devices that will help you see which way the wind blows and measure its speed.

A WEATHER VANE

RATING: ✳ EASY!

You will need
- A wire coat-hanger
- A sheet of paper
- Tape
- A spoon

What to do
1. Lay one end of the loop of the coat-hanger on the paper and draw round it to mark the shape on the paper. Cut out the paper shape and tape it to the hanger, preferably on the end below the point of the hook.

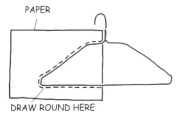

PAPER

DRAW ROUND HERE

2. Hold the handle of the spoon, balance the point of the hook in the spoon, and stand outside in the wind. The hanger will swing round and the end without the paper will point in the direction the wind is coming from.

More things to try
● You can improve your weather vane by adding enough Blu-Tack or paperclips so that the hanger hangs level. Mount the spoon on a support

so that you don't have to hold it yourself – though you can probably work out a more effective mounting system than the spoon!
● If you use plastic or clingfilm instead of paper, your weather vane will survive out in the rain.

A VANE ANEMOMETER

RATING: ✳ EASY!

This will indicate the wind speed.

You will need
- A wire coat-hanger
- Cardboard (e.g. from cereal packet)
- Tape

What to do
1. Bend the loop of the coat-hanger into a square. Cut a piece of cardboard to make a square slightly smaller than the wire square, so that the card will fit through easily.
2. Tape the card so that it hangs from the top of the wire square and can swing easily through it.
3. Hang the apparatus from a convenient branch, washing line or railing, making sure it hangs level by bending the hook if necessary. When the wind blows the card will swing out of the wire square; the more it swings the greater the wind speed.

More things to try
● You can improve your anemometer

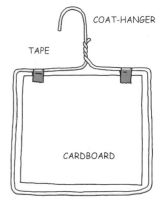

COAT-HANGER

TAPE

CARDBOARD

F M S S M F

by adding a paper scale to show fast, medium and slow wind speeds. You might want to add Blu-Tack or some other weight at the bottom of the wire to prevent the wire from swinging in the wind.
● The main disadvantage of this vane anemometer is that it has to hang at right angles to the wind direction, and when the wind changes you have to turn it round. You might like to design one that can swivel round through 360 degrees with an extra vane that turns it to face the wind. Or you might like to try building a cup anemometer, which solves the problem – see the opposite page.

WHY IT WORKS... Wind is simply moving air. When the wind blows, the moving air pushes harder against the flat paper than against the thin wire. Whichever way the weather vane hangs, a force pushes the paper end downwind. This force is smallest when the paper end points exactly down-wind, and so that is where the vane stops swinging.

A CUP ANEMOMETER

The cup anemometer was invented by Thomas Romney Robinson, Director of Armagh Observatory in Northern Ireland from 1823 until 1882. He calculated that if it could spin with no friction, then the speed at which the cups were spinning would be exactly one-third of the speed of the wind.

You will need
- A sharp knife
- 2 ping-pong balls
- 2 drinking straws
- Glue
- Ballpoint pen
- Bottle-top, thimble or pen cap that fits loosely on the ballpoint

What to do
1. Using the sharp knife, cut the ping-pong balls in half to make four hemispherical cups. To each end of each straw glue one cup so that they point in opposite directions.
2. Make sure the bottle-top, pen cap or thimble swivels easily on the ball-point pen. Tape the straws across the top of the cap or thimble in such a way that all the hollow cups point clockwise. Hold the pen upright (or tape it to a suitable post), balance the cup-and-cap assembly on top,

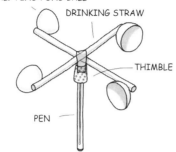

HALF PING-PONG BALL

DRINKING STRAW

THIMBLE

PEN

and your anemometer is complete. The faster the rotor spins, the stronger the wind.

More things to try
● Can you work out how fast your cups are moving? If so, you can measure the wind speed. You may be able to devise a method of fixing a scale to indicate the wind speed.
● You should be able to make your anemometer work by blowing at it from nearby. Can you make it work from a metre away? What is the greatest distance from which you can make it work by blowing?
● Try blowing through a cardboard tube. Does this concentrate your blow? In other words, can you affect the anemometer from further away? Does the length of the tube make any difference? What about its diameter?

A SIMPLE WINDMILL

You will need
- Scissors
- A plastic drinking cup
- A pin
- A piece of wood

What to do
1. Use scissors to make four cuts from the rim to the base of the cup at equal intervals around it. This makes four sails.
2. Bend each one down until it is roughly level with the base, and give each a twist to the left.

3. Push the pin through the centre of the base and into the piece of wood. This pin is the shaft, and your windmill should now spin when you hold it up in the wind.

If you are confident enough to tackle a really ambitious project, try the vertical windmill on the next page.

WHY IT WORKS... When the wind blows, the moving air exerts a force on all the cups, but the force is less when the air can slide over a curved surface and greatest when it goes straight into the hollow cup. So if all the hollow cups point clockwise the rotor spins anti-clockwise.

One great advantage of the cup anemometer is that the instrument does not have to be turned to face the wind; the wind can blow from any direction, and the rotor will always spin the same way.

VERTICAL WINDMILL

Why would you want a vertical windmill? Because all other windmills must be turned to face the wind. Benjamin Wiseman Junior of Diss, Norfolk, solved the problem with his design of 1783.

You will need

- An old bicycle rear wheel, without a tyre but with its axle
- Off-cut of 75mm fence post or similar, for mounting the windmill
- 2 U-bolts and nuts, bolts about 100mm long
- 4 lengths of 12–25mm-square cross-section soft wood, each 60cm, for the masts
- 4 small clamps to hold the masts to the wheel rim
- 8 lengths of dowel, 8–12mm diameter, each about 30cm, for the booms and stays
- Small screw-in hooks and eyes
- String
- Plastic sheet for the sails – black bin bags are fine
- Sticky tape (waterproof if you want the windmill to last)

What to do

1. Although you could make your own pivot and framework, I tend to have a lot of bicycle wheels hanging about, and one of those seemed ideal. Start by mounting the wheel on a piece of 75mm fence post or whatever you're using. You might choose a different piece of timber, depending on how you intend to mount the finished windmill.

2. Mark where you are going to mount the axle of the wheel, and then drill two pairs of holes for the U-bolts; use the bolts to mount the wheel firmly on its base. You can now clamp the assembly for the remainder of the construction – first the spars, then the sails and rigging.

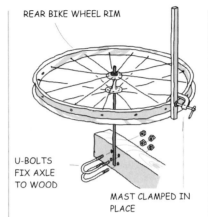

REAR BIKE WHEEL RIM

U-BOLTS FIX AXLE TO WOOD

MAST CLAMPED IN PLACE

Masts and booms

Since this windmill sails round, I'll use sailing terminology. The four masts are made from the 60cm-long square cross-section wood. They are square to make them easy to clamp.

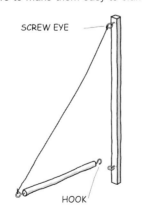

SCREW EYE

HOOK

1. Clamp the four masts evenly round the rim of the wheel, leaving the same side of each mast clear for fixing the booms.

2. The rigging is all going to be fixed with hooks and eyes. Begin by mounting the boom (made from the

dowel): screw an eye into the mast about 25mm above the wheel rim. Screw a hook in one end of the boom and an eye in the other. Use the hook to mount the boom in the eye on the mast.

3. To support the boom, tie a piece of string between the eye at the outer end of the boom, and another eye screwed into the mast just below the top, on the same side as the boom. Tie the string off so that the boom rests in a horizontal position.

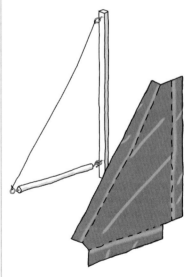

Sails

1. We used black bin-bags, but you may have something more beautiful. Kite nylon would look great. Cut sails slightly larger than the triangle made by the boom, mast and string, leaving 'flaps' on each side. You will need to make a cut-out in the corner where the boom meets the mast, to allow the joint free movement.

2. Wrap the flaps you have made round the mast, boom and string. Use waterproof tape to fix sail in place. Repeat for the remaining sails.

Stays

1. Wiseman realized that the sails need to move up to 45 degrees either side of the straight back position for optimal sailing. To achieve this, use the other four dowels to make stays. These have one screw-eye in the end, and are fixed in line with the spokes between the masts, so that the eye on the end of the stay extends beyond the end of the wheel until it is more or less in line with the end of the boom in front.

2. Tape the stay to the spoke. Tie boom to stay with a piece of string that will limit the travel of the sail to 45 degrees either way – this is your 'main sheet'.

Prepare for sailing

The windmill is now finished, and can be mounted for sailing. You could fix the mounting timber to a fence-post. Don't mount it too high until you have perfected your sailing arrangements – it will need some tweaking.

There are some modern designs of vertical windmill – such as the Sevonius rotor. You'll have seen these outside garages, spinning and alternately displaying first one fluorescent message, then the other. Versions made from cut-in-half oil drums are sometimes used to power irrigation pumps – these don't require huge power and it is more important that the pump keeps going whatever the wind direction. Other designs

involving vertical wings have also been tried, but tend to shake themselves to death.

So, even though the mechanisms of horizontal windmills are complicated and need lots of attention, they're still

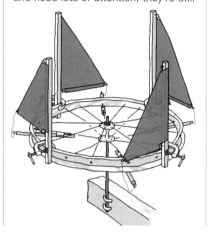

the ones you see grinding corn – because, if you turn them into the wind, then all their sails pull with equal vigour. So, you won't see a Wiseman mill anywhere – unless you build one and put it in your garden!

DID YOU KNOW?

Virtually nothing seems to be known of Wiseman himself. We found him by searching through the Patent Office in London, tipped off by finding his name in a list of Norfolk patents. We can say that he was a merchant from Diss only because that is what it says on his patent of 1783, but there are no further details.

The Patent Office is well worth a visit – the one in London will tell you if there is an outpost near you. But beware: just because an invention is patented, doesn't guarantee it is sensible!

WHY IT WORKS... It is rather surprising that the windmill works at all, given that you can't adjust the sails as they go round. On a boat you would want your sails hauled tight in when you were sailing into the wind, and progressively looser as the wind comes across the boat, and eventually right out like a parachute when the wind is behind. But our windmill does none of that – and still goes round.

Watch it sailing, and you will see that each sail goes through a recognizable cycle – familiar to any sailors. With the wind behind the sail, the boom tends to stick right out as the sail catches the wind. As it turns and the wind comes across the sail, it acts more like an airfoil. This is the most efficient point of sailing and in a boat would be called a reach. I presume the same term applies when sailing a windmill.

As the sail heads round further, it comes into the wind. On a boat, you have to haul the sails in hard to sail close to the wind, and even then you can't ever sail right into the eye of the wind. On our windmill we can't adjust the sails at all, and they flap for about a quarter of the journey. At this point they are a real drag – literally – but fortunately it only happens to one sail at a time, and the others keep on sailing and pull it through. So the power of three very inefficient sails is greater than the drag of one flapping into the wind.

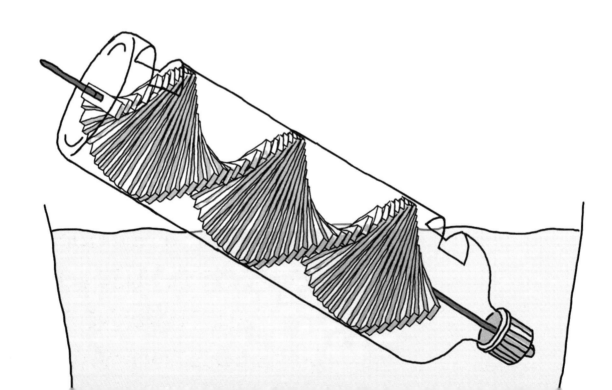

CHAPTER THREE
WATER

ARCHIMEDES' SCREW

Archimedes was perhaps the greatest scientist of the ancient world, and lived mainly in Syracuse, on Sicily, then part of the Greek empire. However, he also worked in Egypt, and when we were there we were amazed to see that this elegant pump is still being used more than 2000 years after he invented it.

You will need

- Empty 1.5-litre plastic drinks bottle – it must have straight sides
- Knitting needle or similar, slightly longer than the bottle
- Lots of wooden lolly sticks, or similar strips of wood. For a standard size bottle you will need up to 120
- Wood glue, such as PVA
- Clear varnish (optional)
- Waterproof tape

What to do

1. Make a hole in the bottle lid and another in the bottom of the bottle that fits the knitting needle tightly. Pass the needle right through the bottle so that a little sticks out of the bottom. Work out how far along the needle from each end the bottle reaches maximum diameter – the lolly sticks are only in the cylindrical part of the bottle.

2. Cut the top off the bottle just below the point where it reaches its maximum diameter. Keep the discarded bit – you will need it later.

3. Cut all the lolly sticks to length so that they just fit inside the bottle; i.e. they're the same length as the bottle diameter. Trim both ends – curved ends won't seal against the bottle sides.

4. Drill each lolly stick exactly though the centre, with a hole the knitting needle will just go through.

5. Now make the screw. First push the knitting needle through the bottle top and into the cut-off portion of the bottle, until it's stopped by the knob at the end. You will be really cross if you make the whole thing and find you haven't done this! Then thread on the lolly sticks.

6. Push the first lolly stick up to the edge of the cut-off bottle top, then thread on the second. Twist the second stick clockwise as far as you can without leaving a gap at the

ends, and glue it in place with PVA. Thread on the third stick, and twist that as far over the second stick as you can without leaving a gap, and so on, gluing as you go.

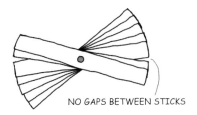

NO GAPS BETWEEN STICKS

7. You are building up a spiral of sticks, and you should get at least three turns within the length of the bottle. When the glue is dry, you may choose to varnish the wood to make it more waterproof.

8. Now it's time to assemble your Archimedes screw. Apply glue along the edge of the spiral, where it will meet the bottle. Now screw the whole thing into the bottle. I suggest doing this so that all the sticks follow the same path as they go down, avoiding smearing glue over the whole of the clear plastic and spoiling the appearance of the finished article. Re-attach the top of the bottle, sealing with waterproof tape. The spike of the knitting needle should emerge from the end of the bottle through the hole you made.

9. When all the glue is dry, cut holes in the bottle for the water to get in and out. These want to be about 5cm across, and extend round about a quarter of the bottle. Position them so that the 'in' hole is just before the

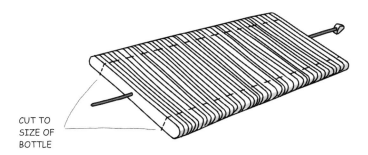

CUT TO SIZE OF BOTTLE

start of the bottom end of the spiral, the 'out' hole just beyond the finish at the top end.

10. Now you can start pumping water. Immerse the lower end of the pump in water, with the body of the pump at an angle of about 30 degrees. Rotate the pump anticlockwise to scoop water into the first turn of the spiral and, as you keep turning, the water is pushed up to the top and out. I'm not sure you can use a little pump like this for anything useful – but at least it will make bath-time more fun.

WHO WAS ARCHIMEDES?

One account says that he may have been the son of an astronomer, another that he came from a poor family. He seems to have been close to the King of Syracuse, Hieron II, who also seems to have had a modest start in life and who tried to encourage Archimedes to use his extraordinary talents for practical purposes.

Archimedes, however, seems to have been obsessed by pure mathematics. Plutarch wrote that he was so 'bewitched by a Siren who always accompanied him, he forgot to nourish himself and omitted to care for his body; and when he was urged by force to bathe and anoint himself, he would still be drawing figures in the ashes or with his finger draw lines on his anointed body'.

So it is a little odd that the most famous story about Archimedes concerns a bath – the one he is supposed to have leaped out of shouting 'Eureka'.

What is really ironic is that the practical feats he is remembered by – which also included a system of leverage to launch a mighty ship, many war machines and the brilliant feat of burning boats in the harbour at Syracuse using mirrors – were despised by Archimedes himself. Indeed, he tried not to write his own theories down at all.

Plutarch says that 'he regarded as ignoble the construction of instruments, and in general every art directed to use and profit, and he only strove after those things which, in their beauty an excellence, remain beyond all contact with the common needs of life'. Apart from an account of his elegant planetariums, he wrote down nothing other than mathematics – and yet his reputation was so great that all these mechanical legends have survived over 2000 years.

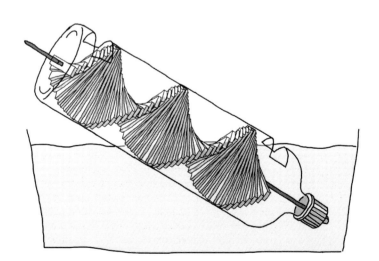

MINI SUBMARINE

Actually, I have no idea whether Cornelus (or Cornelius) Drebbel made a mini-submarine. There are, however, accounts of his full-size one, in which he is supposed to have travelled under the Thames from Westminster to Greenwich and back in the early 17th century.

You will need

- Two plastic fizzy drink bottles, preferably 2 litres each
- A short length of 15mm plastic plumber's pipe (or try clear flexible plastic tubing)
- A plastic, straight-through tap, to fit the above pipe. If you use squashy tube, you could try using a clamp or clothes peg instead
- Some means of fixing the pipe to a bottle top. I used the plastic plumbing inserts that are specially made to fit 15mm pipe (or try the ridged connectors for flexible plastic tubing)
- Lashings of glue – preferably epoxy resin
- Balloon
- A heavy weight, about 2kg (I used a 4lb lead diving weight)
- Gaffer tape
- Coins to adjust ballast (note that only 20p, 5p and 1p coins fit through the bottle neck)

What to do

The aim is to connect the two bottles together head-to-head with plastic plumbing pipe, via a tap. Do the plumbing first. Plumbing used to be rather tricky, involving copper pipe and connectors that needed some skill to make them watertight. However, you can now buy plastic pipe, together with push-fit connectors, which makes the whole business much simpler.

1. Cut a hole in each bottle top for the purpose-made plumbing inserts to pass through. These inserts look like tiny top hats with the top missing, and fit exactly inside the 15mm plastic tube. Pass an insert through the hole in the bottle top, and then into a piece of 15mm tubing about 5cm long. Glue in place with

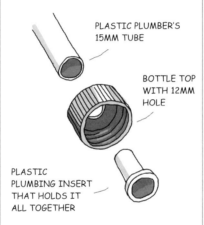

PLASTIC PLUMBER'S 15MM TUBE

BOTTLE TOP WITH 12MM HOLE

PLASTIC PLUMBING INSERT THAT HOLDS IT ALL TOGETHER

a waterproof glue suitable for plastic – I tend to use epoxy, which is a two-part glue you have to mix together and which then sets hard. (You can even get versions that set in five minutes, but they are a bit expensive.)

2. Now fix the tap between the two pipes, connecting the bottle tops. If you have bought a plastic tap, it will come with fittings that let you push the 15mm pipe into it for a watertight seal. If you don't want to splash out, you could make your own tap. Use a piece of floppy tubing such as clear PVC tube for home brewing, or even a bit of hose pipe about 7.5cm long. It will need to be a size to fit tightly over the plumbing pipe – or even directly over the inserts, missing out the

plumbing pipe altogether. You could then use a small clamp to squash the tube closed, instead of a tap.

3. Now prepare the bottles. One you're just going to fill with water. The other is the air reservoir. First install the balloon. Push it into the bottle, and then stretch the neck of the balloon over the thread of the bottle.

4. Next, adjust the buoyancy so that it just floats: submerge the bottle so that the balloon just fills with water. You need to attach a weight to the outside – I strapped mine on with gaffer tape.

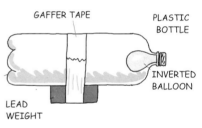

GAFFER TAPE

PLASTIC BOTTLE

INVERTED BALLOON

LEAD WEIGHT

5. If you need to adjust the weight further, you could add coins to the inside of the bottle. You may have to readjust when the whole sub is assembled.

6. When you are ready, attach the bottles to the plumbing. This should be done in the bath or swimming pool, because the unsupported water-filled bottles could damage the fittings. I found that the stretched neck of the balloon sometimes ripped when you screwed the bottle top over it, so try binding it with sticky tape first. Screw the air reservoir bottle on; your sub should still float.

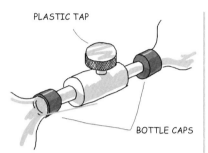

PLASTIC TAP

BOTTLE CAPS

Running the submarine

1. Now the exciting bit. The second bottle is really just used as a water-pump. Screw it on to the free bottle top. Open the tap. Putting the water-bottle under your arm, squeeze the bottle so that the balloon in the other bottle fills with water. Close the tap. Unscrew the water-bottle. Release the submarine and it should sink. If it doesn't, despite the balloon being filled with water, try adding more weight.

2. So far so good. But when you are under water, with no connection to the air above, how do you get rid of the excess water, fill your tanks with air, and rise to the surface? Simple – open the tap! The balloon deflates, and the sub rises to the surface.

More things to try

● You don't need to detach the water-reservoir when the balloon is filled with water. At first this seems wrong – surely you are moving weight (in the form of water) from one bottle to another and, if you leave the water-bottle attached, the weight of the complete sub has not changed?

True, but the volume has. You squashed the water bottle, and then squashed the air into a smaller space to make room for the transferred water. So the complete two-bottle sub weighs the same as it did, but takes up a smaller volume. If you adjusted it so that the whole thing only just floated, then it must now sink, because it displaces less water and so gets less 'push' from the water – Archimedes' Law in action!

CORNELUS DREBBEL

Drebbel is surrounded by so much myth that it is sometimes hard to know what he really did and what is just legend. He certainly seems to have come to England around 1605, invited by King James I. On the silly

side, he is supposed to have made a machine that produced lightning, rain, thunder and extreme cold – which, when demonstrated at Westminster, drove the people out into the street. He was also responsible for an instrument for 'showing pictures of people not present'. These sound like theatrical special effects.

The submarine, however, seems more real. He apparently used pig bladders instead of balloons, and the craft was powered by oars passing through leather seals. We tried rowing underwater and can report that it is possible – just. Clearly, sinking a boat is something any old fool can do. Drebbel may have been the first to make one come up again.

Most surprising is an account of a bottle he had on board that, when he opened it, could refresh the stale air. Presumably it either got rid of carbon dioxide, or added oxygen. I say that it is surprising because oxygen and carbon dioxide were not discovered until at least 150 years after Drebbel's amazing underwater voyage!

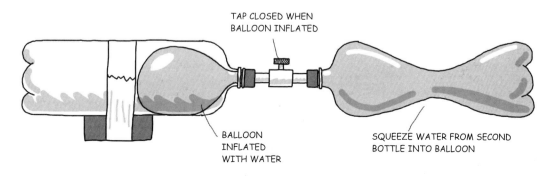

TAP CLOSED WHEN
BALLOON INFLATED

BALLOON
INFLATED
WITH WATER

SQUEEZE WATER FROM SECOND
BOTTLE INTO BALLOON

GOING WITH THE FLOW

You might have used a siphon yourself, cleaning out a fish tank, perhaps – putting the short arm of the tube under the water, and getting the flow going by sucking the other end of the tube (and hoping not to choke on fish muck!).

SIMPLE SIPHON

RATING: ✱ EASY!

All you need is a tank filled with water and a tube – a piece of hosepipe will do – in the shape of an upside-down U, with one long arm and one short arm. Fill the tube with liquid and put the short arm in the tank; liquid will then flow out through the tube as long as the short arm stays beneath the surface and the end of the long arm is below this level. You can empty the whole tank in this way.

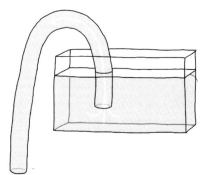

AS LONG AS THIS END OF THE TUBE IS BELOW THE SURFACE OF THE LIQUID IN THE TANK, THE LIQUID WILL FLOW OUT THROUGH THE TUBE.

AUTOMATIC SIPHON

RATING: ✱✱✱ CHALLENGING

What you need

- An old pop bottle, any size from 1 to 3 litres (make sure you keep the screw top)
- A plastic tank connector to fit a tube about 2–3cm diameter
- A soft plastic tube, 40cm long and the right diameter to make a good push fit on the tank connector
- Glue, if necessary
- A flat plastic lid, with diameter a bit less than that of the bottle
- A floppy plastic sheet (e.g. from an old document folder)
- A piece of stiff wire at least as long as the bottle (a bicycle spoke is ideal)
- A nail or paper-clip
- A 'tank' at least as deep as the bottle up to the tank connector (e.g. fishtank, washing-up bowl or large saucepan)
- Some flexible wire

What to do

1. Drill or cut a hole high on the shoulder of the bottle, just where the straight sides start to curve, to fit the tank connector.

2. Using a pair of scissors, cut the bottom off the pop bottle, at the bottom of the straight sides and above the tough curved base.

3. Fit the tank connector to the bottle. Fit the plastic tube to the tank connector, using glue if necessary.

4. Make a small hole (for the wire) in the centre of your flat lid. Make several large holes (we drilled 2.5cm holes) in the lid, all around the centre, to leave just a skeleton.

5. Cut a disc of the floppy plastic sheet to fit neatly inside the bottle, ideally just touching all round. Make a small hole in the centre of this floppy disc.

6. Make a small hole in the screw top of the pop bottle. It's important that this hole is only just big enough for the wire to go through. Screw the top on.

7. Pass the wire through the screw top, the bottle, the floppy disc, and then the skeleton lid, and tie the nail or paper clip across the end with wire so that the lid cannot slip off the bottom. Your siphon is now ready to use.

8. To get it working properly you need a tank or basin at least as deep as the bottle up to the tank connector. If you haven't got a spare fish tank, try a washing-up bowl or a deep saucepan.

9. Fill the tank with water, and lower the whole siphon slowly in. The water should lift the floppy plastic off the flat lid and fill the pop bottle up to the side arm. The floppy plastic should then drop back on to the lid.

10. Make sure the plastic tube is pointing into a bucket or drain or somewhere that won't make a terrible mess.

11. Hold the bottle by the neck or side arm, and pull up the top of the wire or spoke. Give it a steady pull,

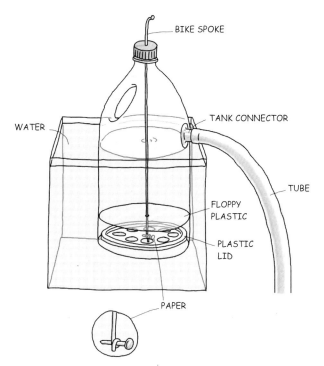

BIKE SPOKE

TANK CONNECTOR

WATER

TUBE

FLOPPY
PLASTIC

PLASTIC
LID

PAPER

so that the water in the bottle is lifted up and begins to flow through the side arm. This is quite tricky, but when you get the hang of it the siphon should take over, and will empty the entire tank in one smooth flow. And with this automatic system you don't have to start by sucking!

More things to try
● Almost every lavatory cistern in Britain is fitted with a siphon. The cistern is the water tank, usually above and behind the lavatory bowl. Most cisterns have loose lids that can be lifted off, and then you can see inside. Why not have a look in your cistern?

The inlet pipe is usually in one top corner, and the flow of water is controlled by a valve, often a float valve or ballcock – a ball of plastic or sometimes metal floats on the water and, as the water level rises, its long arm is an effective lever to shut off the water supply.

The siphon is usually in the middle of the cistern, and to see how it works flush the loo while you are watching. The movement of the piston inside is just like the one in the diagram.

DID YOU KNOW?
Siphons were introduced into lavatory cisterns in the 1850s, when the water boards were worried about leaky valves dribbling into lavatories all day and all night and wasting masses of water. The authorities insisted on siphons, because with a siphonic flush the cistern cannot leak; there is no valve letting the water out of the bottom of the cistern. For this reason siphons were called 'valveless water-waste preventers'.

In other countries the water authorities seem not to have worried about wasting water, and siphons are less common. The typical American cistern has a 'flapper valve' – a simple flap of rubber over the outlet pipe in the bottom. Pressing the flushing lever lifts the flapper valve and allows all the water to run out.

So, next time you go abroad, go to the loo, take the top off the cistern if you can, flush, and watch. Then please write and tell me about it – even better send a photo or a sketch!

WHY IT WORKS... The liquid has to flow uphill to get out of the tank, and at first sight this is surprising. However, it will do so because it is finding a lower level. The weight of liquid is greater in the longer arm – the arm outside the tank – than in the shorter arm; so it tends to run out, and atmospheric pressure forces the liquid in the tank up the tube. If bubbles of air get into the tube the liquid column can break, and the siphon stops working.

ALL CISTERNS GO!

This cistern is based on a design by Joseph Adamson, a plumber in Leeds. In 1853 he took out what I believe was the first British patent that included a siphon in the cistern. He clearly intended it for use in factories and other places for regular flushing of urinals.

You will need

- A plastic fish tank or similar box, about 30cm x 15cm x 15cm
- 40cm of slightly flexible plastic tube, about 2.5–3cm diameter
- Epoxy or other waterproof glue such as bathroom mastic
- A hinged support to hold the platform about 5cm above the base (see below)
- 50cm of sticky-backed Velcro
- A wooden base (e.g. 20mm softwood) and a wooden platform (e.g. 4mm ply), each about 40cm x 15cm
- A weight of about 1kg

What to do

1. Drill a hole in one end of the fish tank to fit your plastic tube. My tube has an outside diameter of 3cm, and I drilled a 3cm hole with its centre 5cm below the top of the tank.

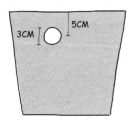

I found it best to do this slowly, using a firm wooden support below the plastic, and drilling right into the wood; then the plastic did not crack.

2. Cut about 12cm off the end of your plastic pipe, at an angle of 45 degrees. Then you can join the pieces together to make a right-angled pipe.

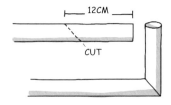

3. Glue the two pieces together to make the right-angled pipe.

4. When the glue is dry, feed the long end through the hole in the end of the tank from the inside, twist the tube until the inside short end is roughly vertical, and then mark it about 1cm from the bottom of the tank.

5. Remove the pipe, trim off the bottom end, replace the pipe and glue it in place – see finished cistern opposite.

6. Next make the hinged support. I used two old brass brackets screwed to the sides of the base, but you could use wire, or plywood. For an axle or spindle I used an old knitting needle, but a long nail would do, or a Meccano rod. I fixed my spindle to my platform with two electrical staples hammered into a piece of scrap wood glued underneath.

This hinged support should be closer to one end than the other – my hinge is just 15cm from one end of the base and platform.

7. Next, fix the fish tank to the platform. I used sticky-backed Velcro for this, because it is strong and flexible. Stick two strips of Velcro to the underside of the tank and two matching ones to the platform. My strips were about 20cm long, and about 3cm to the long side of the spindle.

8. Finally, fix the weight to the platform. Again I used twin strips of Velcro and fixed the weight with its centre about 3cm from the end of the platform – but you will have to find the correct position by trial and error!

Now your cistern is complete.

Making the cistern work

1. When you stick the weight on, the whole platform will tilt back to the left – the short end. Now run water in slowly, from a tap or hose. As the tank fills, the balance of weight shifts towards the right, because most of the tank is to the right of the pivot.

2. If you have placed both the tank

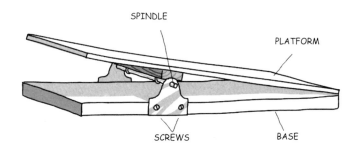

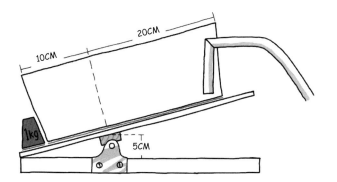

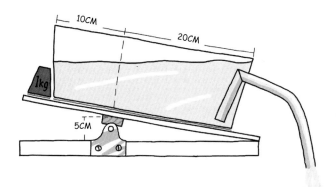

The successors to these Adamson tippers can be found in most communal gents' urinals, in pubs, factories and offices. The unit is usually enclosed, so that you can't see what is going on. Probably they only tip the contents out, without a siphon. However, the principle of tipping at regular intervals dates back to Joseph Adamson's patent of 1853.

After Adamson's pioneering idea, siphons became immensely fashionable; they did away with leaky valves, and were called valveless water-waste preventers. Soon the authorities insisted that all cisterns should be fitted with siphons, and even today British Standard 7357 (1990) requires that 'cisterns shall be supplied with an efficient flushing apparatus of the valveless siphonic type which prevents the waste of water'.

In 1854, the flamboyant plumber Josiah George Jennings patented a lavatory-and-cistern combination that is almost identical to today's; even though hundreds more 'improvements' were made in the second half of the 19th century, none of the innovations has lasted. Jennings got it right.

Some people think that the flushing toilet was invented by Thomas Crapper, but in fact he was born too late. He started his plumbing business in 1861, after the important inventions had been made. Crapper was a competent plumber, but not a great inventor; he took out nine patents, but none of them was memorable.

and the weight in the right positions, when the water is about 1cm from overflowing, the whole tank will topple to the right and (with luck) the tube will fill and the entire contents will siphon out with a good strong flush.
3. When the flush is finished, the tank will tilt back to the left again, and start filling once more. By adjusting the flow of the incoming water, you can make the tank flush every minute or every twenty minutes, as you wish.

4. If the tank overflows before it tips, try moving first the weight and then the tank to the right, a few milli-metres at a time.

If the tank tips but does not siphon properly, try moving first the weight and then the tank to the left, a few millimetres at a time.

With a little practice you should be able to get it to work every time – although mine is a little temper-amental! – and flush with pride!

MAKE SOLITARY WAVES

These strange waves were discovered by John Scott Russell. They can bounce, pass through each other, and travel huge distances. They are very easy to make – even in the bath. But these waves are really fun if you can give them a decent length of run.

You might want to construct some simple apparatus to investigate them in more detail.

The most important thing is the height of the wave in relation to the depth of water: too shallow and it will disperse, too steep and it will break. Solitary waves seem to settle into an even state, not breaking or dispersing.

In practice it's easier than it sounds, so don't worry.

You will need
- A suitable wave tank (see below for ready-made or DIY)
- A paddle

You want a tank at least 2 metres long and about 10cm across and deep. If you are ambitious you could build a really long tank – I'd like to know just how far you can get solitary waves to travel in a home-made tank. For tank construction, I suggest a bit of guttering.

You can get square-section guttering, which I prefer to the usual U-shape because it has even depth over the whole width. Most big DIY stores sell systems including suitable end caps, which can be used to seal the ends of the tank and stop the water falling out. Some of these are self-sealing with proper rubber joints, others may need sealing with silicone bathroom sealant. Or you could make wooden end caps.

If you go for the wooden option, make them square and bigger than the guttering so that they act as a support and stop it toppling over.

MAKING YOUR OWN TANK

You will need
- Coated chipboard for the back, bottom and ends of the tank – the sort sold in DIY stores for shelving is good. It's easy to work with, and the coating is reasonably waterproof
- Chipboard screws
- Wooden beading 12mm square: one piece as long as the tank, and one for each end of the clear plastic
- 19mm nails to fix beading
- Acrylic double glazing sheet
- Silicone bathroom sealant

In the show, I had a plastic see-through tank and you could make one from Perspex (rather expensive) or another see-through plastic. The stuff used for double-glazing isn't bad – though rather hard to cut cleanly. You have to score it with a knife, and then snap the sheet to break it. Practise on an off-cut first.

What to do
1. Make the chipboard box first, fixing the back, bottom and ends with chipboard screws, then nail on the beading, and slip the acrylic sheet inside. Seal everything with silicone bathroom sealant.

If you don't mind a slightly shorter tank, the clear plastic cover from a fluorescent light fitting might make a nice ready-made trough.

Making waves
1. Fill the tank half full of water. The easiest way to make waves is with the paddle – a piece of thin wood or plastic that just fits in the tank.
2. Dip it in the tank at one end until it touches the bottom, and give the water a shove – piling it up in front of the paddle.

Solitary waves are supposed to form when the wave height (I think this means height above the water) is comparable to the depth of the water.

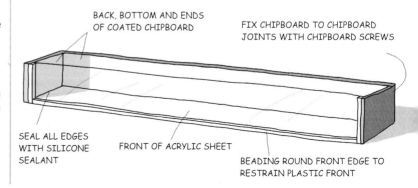

BACK, BOTTOM AND ENDS OF COATED CHIPBOARD

FIX CHIPBOARD TO CHIPBOARD JOINTS WITH CHIPBOARD SCREWS

SEAL ALL EDGES WITH SILICONE SEALANT

FRONT OF ACRYLIC SHEET

BEADING ROUND FRONT EDGE TO RESTRAIN PLASTIC FRONT

I remember seeing a huge – 10-metre-long – soliton (as a solitary wave is now known) tank at York University that used a more professional mechanism: a sluice. There was a tight-fitting gate near one end of the tank, and the section behind was filled to a greater depth than the rest of the tank. As the gate was raised, a pulse of water shot down the tank. This was great for making sure you could repeat particular waves exactly – but it may be a bit over the top for your purposes.

Nevertheless, you should practise getting waves of consistent size, because the height of a solitary wave affects its speed.

More things to try
● Distance: how far will a single wave travel?
● Bounce: how many bounces before the wave dissipates?
● Overlapping: what happens when one wave overtakes another?
● Splitting: waves that are 'too big' may split into two.
● Speed: what affects how fast waves travel?

DID YOU KNOW?

John Scott Russell had grown up in Glasgow where he was so fascinated by the great roar of the first Newcomen steam engines at the Carntyne mines that he abandoned his career in the church to become an engineer himself.

Russell accepted an invitation from the Union Canal Company to beat off the challenge from the new steam carriages and railways by designing better, faster boats – this was the turning point in his career. While testing his boats on the Grand Union Canal near Edinburgh, he decided that it was the great bow wave the boats made that was slowing them down.

As he rode along the canal in August 1834, he watched a rapidly drawn boat as it suddenly came to a halt in front of him. And something extraordinary happened: the great hump of water built up in front of the boat kept on moving as a single, huge wave, apparently without losing speed.

Russell set off on horseback to follow this wave, and chased it for over a mile along the canal before it started to weaken.

This was no ordinary wave – Russell knew that Bernouilli and Newton had described exactly how waves travel and disperse, but this one didn't follow any of the rules, it just kept going. He was convinced he'd seen something special and set out to discover what it was in a series of elegant experiments.

He'd been right about the bow wave slowing canal boats down, but found out that by pulling the boat at just the right speed it could rise up on to this 'wave of translation' and surf comfortably along with very little effort. On the strength of this, he introduced a new night sleeper canal service from Edinburgh to Glasgow, and found that the horses could easily keep going if he kept the speed just right.

Solitary waves at sea don't always have the desired effect. Recently there have been disturbing reports of fishing boats being suddenly overwhelmed by huge single waves coming out of the blue. Initial theories (not yet confirmed) suggest that these are solitary waves produced when new designs of high-speed boat hit the critical combination of boat speed and water depth.

It wasn't until the 1960s that scientists realized just how important Russell's discovery had been. They reasoned that if a wave in water could be made to travel so far, what about other waves, such as light?

Today's most advanced fibre-optic communications use stable pulses of light identical to Russell's solitary waves (now called solitons), to carry masses of information over thousands of kilometres of fibres.

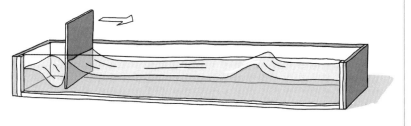

CHAPTER FOUR
BIOLOGY

CHROMATOGRAPHY

Biologists who want to find out about the chemistry of living things often need to separate out the molecules that things are made of. One cunning technique is chromatography – here is a very simple version.

SIMPLE CHROMATOGRAPHY

RATING: ✳ EASY!

You will need
- Blotting paper or (better) filter paper cut into strips 2.5 x 30cm
- A tall glass, e.g. a pint beer glass
- Water
- A test specimen (see below) – start with a brown Smartie

What to do
1. Put the paper strips into the dry glass and fold them over the edge so that they hang nicely and just reach the bottom. You need strips longer than the glass because as the paper gets wet and heavy there has to be enough paper overhanging to stop it slipping down.

2. Remove the strips from the glass, and mark a faint pencil line about 3.5cm from the lower end of each strip. Put 2.5cm of water in the bottom of the glass.

BLOTTING PAPER FOLDED OVER GLASS

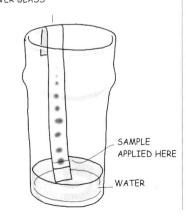

SAMPLE APPLIED HERE

WATER

3. Prepare your specimen. A brown Smartie works rather well; moisten the surface of the sweet, and rub the colour on to the surface of one of the strips of paper just above the pencil line. Try to keep the colour in a neat spot.

4. Lower the strip into the glass, specimen down, so that it hangs over the edge as before, but with the lower end in the water.

5. Wait about 20 minutes, as water soaks up the paper, taking the food colouring with it. You can then dry the paper carefully. Take care, though: if you hang it directly on a radiator you may transfer some of the colour to the radiator.

6. What happened? Is the brown dye made up of just one substance, or many?

More things to try
● You can try all sorts of coloured specimens – ink, felt-tip pen, paint, food colouring, coffee, tea. Also naturally coloured food such as red cabbage, carrot juice, squashed-up pea juice.

● For fruit and veg, you have to extract some of the colour into liquid. Chop the sample finely, then grind it up with some water. If you have a juicer, use it. Next choice would be a pestle and mortar, or a blender. Or you could force it through a sieve. Use as little water as possible, because you want the extract to be as concentrated as you can get it.

● Put a few drops of the now coloured liquid on to the pencil line you marked on a paper strip. Try to keep the blob on the paper as small as possible: it's going to spread as it goes and the more spread-out it is to start with, the harder it will be to see separate lines.

● You will find that, despite grinding, you can't get some colours out of the sample and into solution. This may be because they are not water-soluble. Most cells have a watery bit and a fatty bit, and whether a particular substance will dissolve in water depends upon where it normally lives in a cell.

● If you want to experiment with colours that don't dissolve in water, try using different solvents.

DIFFERENT SOLVENTS

RATING: ✳✳ MODERATE

DANGER FROM FUMES; SOME SOLVENTS FLAMMABLE

You may find that some substances don't move much using water because they don't dissolve readily in it. In other cases you may find that the coloured patch moves, but doesn't separate. In this case, why not try a different solvent? Because some of these solvents have unpleasant fumes, use a slightly different set-up with smaller quantities and a lid to contain the fumes.

You will need
- Blotting or filter paper cut into strips 1 x 10cm
- A small glass
- A tall glass, e.g. a pint beer glass, to cover the small glass to keep fumes in

- A test specimen (see below)
- Solvents (see below)

This experiment is particularly suited to biological substances. You could try grass or leaves, and vegetables such as tomatoes, red cabbage or carrots.

Solvents: as well as water, there is cooking oil, acetone (nail varnish remover), alcohol, methylated spirits, dilute washing-up liquid, and many others you could try, but please be careful and read any safety instructions first.

Note: a solvent is simply the liquid (usually) in which something is dissolved. The thing dissolved in it is called a solute.

What to do

1. The set-up is simply the previous experiment, but on a smaller scale and with a lid. Put the paper strips into the glass, and bend over the rim. Remove them and mark a faint pencil line about 1cm from the bottom. Pour only 1cm depth of the liquid (solvent) you are using into the bottom of the glass.

2. Prepare your specimen. Do by all means try Smarties again, but it's fun to try something a bit more challenging. If you've decided to use a particular solvent, you should also use it to prepare your specimen. So if you want to try the red colour in tomatoes with cooking oil, you ought to grind up your tomatoes with cooking oil.

3. Apply a small patch of your specimen to the pencil line as before. It's a good idea to get rid of as much solid matter as you possibly can, because this will just result in a streak as more colour dissolves.

4. Put the paper strip with the specimen into the glass, with the bottom in your solvent of choice. Put the pint glass over the top of the whole thing, as a lid. You don't have to do this if you are using, say, cooking oil, but it is good practice if you are using a solvent such as acetone that produces flammable or unpleasant fumes. Wait 20 minutes or more as the chromatography progresses.

5. Take care where you do this. In particular, if you are performing in the kitchen, KEEP AWAY FROM NAKED FLAMES. Some solvents produce flammable fumes that are heavier than air, and could 'creep' along a work surface to the cooker. Most of these solvents are household items – but it makes sense to take care.

6. When chromatography is complete, you will want to examine and perhaps sketch your result. If you want to preserve the paper, dry it carefully – not on a radiator but outside in the breeze. Some substances – especially in plants – are not visible in daylight, but do show up in ultra-violet light. To see them, why not take your results to the disco!

RICHARD SYNGE

Being able to separate substances is very important in chemistry – even today in Holland, chemistry is called 'Scheikunde', or the 'art of separation'.

Richard Synge, who worked at the wool industry research station in Leeds, thought he knew how to make chromatography a much more sensitive process. Rather than using paper with a liquid moving over it, he used two liquids. By fine-tuning the system he found he could separate very similar molecules. Synge was well known for his eccentric dress-sense and love of trainspotting. But his simple system proved incredibly powerful, so much so that he was awarded the Nobel Prize for Chemistry in 1952.

WHY IT WORKS… When we say that sugar dissolves in water, it's only part of the story. Different things dissolve in water to a different degree. So if you can dissolve one teaspoon of sugar in a glass of water, you will be able to dissolve much more salt.

The chromatography described here is 'liquid paper' chromatography. You apply your sample to the paper, and then liquid flows over it. Each substance in your sample will be more or less likely to dissolve in the liquid. It will also be more or less likely to stick to the paper. Chromatography depends upon the balance of the two. Particles of coloured stuff move by jumping into the liquid, then sticking on to the paper and so on. If they spend more time in the liquid, they move quickly. But if they are more attracted to the paper, and spend more time there, they move slowly.

KIWI FRUIT DNA

What you look like, how your body is put together, and what your children will be like is largely controlled by the information coded in DNA. I always thought of DNA as some kind of mysterious substance buried away in our cells never to be seen – but it isn't. Here is a recipe to reveal DNA in your own kitchen.

You will need

- Bottle of methylated spirits
- A large saucepan full of ice
- Salt (the kind you put on food)
- Washing-up liquid (buy the cheapest you can find. The concentrated type doesn't work as well – a supermarket's own brand is perfect)
- Kitchen scales
- A medicine spoon (5ml)
- A measuring jug
- Kiwi fruit (a ripe one works best)
- Knife and a chopping board
- A small bowl
- A large saucepan full of hot water (not boiling – about 60 degrees, just as it comes out of the hot tap)
- Coffee filter and filter funnel
- A champagne flute (or any tall skinny glass)
- A bit of wire (fuse wire is good)

What to do

1. Push the bottle of methylated spirits in the ice, to get it cooling straightaway. IT IS NOT CONSIDERED SAFE TO PUT METHS OR ANYTHING INFLAMMABLE INTO THE FRIDGE OR FREEZER. They are electrical devices and, although it is unlikely, a spark could ignite fumes.

2. You need to make up a solution of 3g of salt, 10ml of washing-up liquid (detergent) and 100ml of water. Measuring just 3g of salt is not easy. You may find that your scales will only weigh out a minimum of 25g.

3. If this is the case, you could simply make up more and discard the excess: use the 25g of salt, add 80ml of washing up liquid (16 measuring spoons) and make up to just less than one litre. More elegantly, using the 5ml medicine spoon, add salt to the scales, counting the number of spoonfuls of salt in 30g or 60g, then divide by 10 or 20 to get the number of spoonfuls

KIWI

CHOP FINELY

ADD TO SALT, WATER, DETERGENT MIX

SIT THE BOWL IN THE SAUCEPAN OF HOT WATER FOR 15 MINUTES. MAKE SURE THE WATER CAN'T GET INTO THE MIXTURE.

in 3g. Either way, stir your mixture thoroughly (avoid froth) to dissolve the salt.

Prepare the kiwi fruit

1. Peel one kiwi fruit and chop into little bits. Scoop the bits into the small bowl and add about 100ml of the salt–detergent mix.

2. Sit the bowl in the saucepan of hot water and leave for 15 minutes. Make sure the water can't actually get into the mixture.

PASS THROUGH A FILTER INTO THE WINE GLASS

3. Pour the green mush into the coffee filter, and catch the liquid that comes out in the champagne flute. You will need about one-fifth of a glass. This is now a solution of various bits from the broken-down cells – including DNA.

Get the DNA out of the solution

1. Very carefully drizzle the ice-cold meths down the inside of the champagne flute so that it forms a purple layer on top of the green

layer. When you have added a layer of meths equal to the layer of green extract, set the glass on the table and watch.

2. You should almost immediately see a white layer beginning to form at the boundary between the green and the

AIR BUBBLES LIFTING DNA

METHYLATED SPIRITS

STRANDS OF DNA

KIWI MUSH

CHAMPAGNE FLUTE

purple. If you look carefully you may see that the layer is made up of filaments – DNA from kiwi fruit in your kitchen.

Pull the DNA out

1. The DNA is visible because it has long molecules that tend to get tangled together, and it is this tangle that you are pulling out of the glass.

2. Make a small loop – 5mm – at the end of the wire, and use this to carefully pull out the DNA. It's whitish purple because it's still contaminated with meths. But this is it: the molecule of life.

WHY IT WORKS... DNA is found in every cell of every living thing, including kiwi fruit. To get it out, you have to use a series of biochemical tricks. You then have to disentangle the DNA from the protein and all the other stuff that is released at the same time.

The first part of the problem is solved by chopping the kiwi fruit up and letting it soak in detergent and salt. The detergent strips away the cell membranes (that hold in the insides of the cell) and allows the internal goo to escape. The finer you chop your kiwi, the more cells get broken open – however, don't try liquidizing it or you will smash the DNA to bits.

Getting rid of the protein that sticks to the DNA is all done for you. Kiwi fruit naturally contains lots of a special enzyme called a proteinase. Enzymes are like small molecular machines and they can do a variety of clever things (the enzymes in washing powder digest fat, for example). The proteinase enzyme attacks the proteins clinging to the DNA and breaks them up, releasing the DNA. Only kiwi fruit has enough of these enzymes already there – so don't try an apple or orange.

The green watery layer in your champagne flute is full of DNA as well as lots of broken-up proteins and a load of other stuff. When you pour the cold meths on to the green layer, you make the DNA dissolved in the water turn into a solid as it can't remain in the meths. You often get little bubbles forming at the boundary that float strands of DNA up into the meths – the bubbles are probably caused by the temperature difference between the layers making the air dissolved in the green layer come out of the solution.

WEIGHING THE EVIDENCE

A seed contains a small amount of food for when the new plant starts growing, but it obviously can't provide all the nourishment needed for a lifetime: little acorns can't feed great oaks all by themselves. So just where does the stuff that plants are made of come from? This demonstration, using common-or-garden cress, may help you to find out.

You will need

- Block of wood, approx 2.5 x 2.5 x 1cm
- Card for the base, 15cm square
- Glue
- 2 strips of stiff card, each 10 x 2.5cm
- 2 plastic lids, e.g. the clear ones on tubs of cream
- Pins
- Plastic drinking straw
- Cotton thread
- Blotting paper — enough for a circle to fill each lid
- Cress seeds
- Some grains of sand (see below)
- Water dropper

What to do

1. You're going to make a balance with the cress growing on just one side. First make the body of the balance. Glue the block of wood on to the centre of the base, then glue the cardboard strips vertically on to each side. It is important that the tops of the strips are at the same level.

2. Now make the balance pans from the plastic lids. Use a pin to make three holes, evenly spaced, around the rim of the plastic lid. Make a hole in each end of the straw. Use the cotton thread to suspend a lid from each end of the straw.

3. Suspend the straw by pushing a pin through the middle of it, and resting it on the top of the cardboard strips. Try to get the balance as level as possible, but don't worry if it isn't absolutely right.

4. Fold the blotting paper in half to allow you to cut a pair of identical circles to fit inside the lids. Put the circles in each lid.

Growing the cress

1. Cress, like many plants, will grow better in fairly warm, light conditions. A windowsill is good in summer; one above a radiator or at least free of draughts should be used in winter. Move the balance to the growing site now – it will be a bit tricky to move once it contains seeds and water.

2. Scatter cress seeds over the blotting paper in one balance pan only. You will want a reasonable crop, so be fairly generous, although you shouldn't actually pile them up! Leave a little blotting paper visible round the edge so that you can see whether the cress is drying out.

3. This will upset the balance, of course, so add some sand or similar – you need something that doesn't dissolve in water – to the other pan until the balance is level once more.

4. Now water the cress. The first watering is easy – just add water until the blotting paper is fairly wet on the cress side of the balance, and then add an identical amount of water to the other side to bring the balance level once again.

5. These are the starting conditions for the experiment: cress and water on one side, sand and water on the other and the balance level.

6. The challenge is to keep the experiment controlled so that you can be quite sure what (if anything) is causing the balance to tip.

Keeping the cress wet

Cress takes a few days to germinate, and will grow quite happily over a fortnight. However, during this time the blotting paper may dry out. Your problem is to make sure that the amount of water you add does not in itself upset the balance, so use the water dropper to add equal amounts of water to each side. If you have to add five drops to the cress, add five drops to the sand.

The idea is NOT to bring the balance back to the level each time, but to restore it to the position it was in just before the watering.

The results

What happens? As the cress grows, does it get heavier? Since each side of the balance weighed the same before the experiment, and received identical amounts of water, where did the extra weight come from?

Is it the water?

Most living things contain a lot of water, so it is possible that the extra weight in your cress is in fact the water you added. Because there was no cress on the other side of the balance, the water simply evaporated. The cress trapped its water.

Testing the water theory

You could test this idea by getting rid of the water. You should stop watering, and allow the cress and the sand to dry out. This will work much better if you put it in an airing cupboard until it is completely dry.

Is the cress still heavier or not?

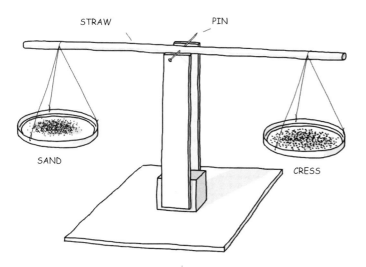

STRAW   PIN

SAND

CRESS

More things to try

● This is not a perfect experiment. It would be nice to eliminate evaporation from the blotting paper altogether, but this is tricky with cress. You could repeat the experiment using, say, runner beans – just use one. At the beginning, add water to the lids and then cover with clingfilm. When the plant starts to grow, make a tiny hole to let it through and then seal the system again, perhaps using a little Vaseline.

● Another problem with this experiment is that plants themselves actually move water. They take in water through the roots, but lose it from the leaves – this is called transpiration. So in some circumstances the cress can actually lose weight for a time because it loses water by evaporation, like the sand, but also by transpiration. This is why it is important to dry the system out at the end, so that you can 'weigh' the actual stuff the cress is made of, eliminating the water.

ADAM SAYS

Recently, at an American university, graduating biologists were presented with a log of wood and invited to explain where the stuff that makes up wood came from. Most (and remember these were people with very recent biology degrees) couldn't bring themselves to believe that the molecules in the wood come from the air (see below for explanation). After all, it just doesn't seem right, does it?

It didn't seem right 300 years ago when Stephen Hales, a vicar from Teddington, first tried experiments like these (and see overleaf). But that was the very reason he was doing the experiments in the first place. Science deals in the things that don't make common or any other sense, and tries to make sense of them. This is also why really revolutionary science often has a hard time being accepted.

WHY IT WORKS... It was thought for a long time that a plant's 'food' comes from the soil – that nutrients are drawn up through the roots. Indeed, some minerals are drawn up along with water, and are useful to the plant.

However, when the cress grows, the blotting paper does not disappear. You could grow many plants on blotting paper or the equivalent, and provided you gave them enough water together with the minerals they need, they would grow quite happily. No doubt you will have grown beans in jam jars with blotting paper at school. But the plants are not made of minerals, or blotting paper. Plants, like most living things, are made from substances based on carbon – things like sugars, starches, cellulose and so on. We have not added sugar, so the plant must have made it – but what is the raw material?

The amazing answer is: thin air. Plants use the energy of sunlight to manufacture sugars from air. Given water, carbon dioxide and sunlight, plants use photosynthesis to make sugar, which is food and also the raw material for all the other things plants are made of. So when you pick up a great heavy log, remember that it is made of thin air because all the carbon in it came from carbon dioxide in the air.

HOW SAP RISES

Stephen Hales (1677–1761), Vicar of Teddington in Middlesex, was the first to work out how sap gets up a plant. He thought there were three processes, in the leaves, the roots and the stems – and these are the same for every plant, whether a tiny daisy or a towering tree.

THE LEAVES

You will need
- Food colouring (red is good)
- 2 small glasses
- Sharp knife
- 2 stalks of fresh celery, with leaves
- Hairdryer (optional)

What to do
1. Dilute food colouring with water – you want a mixture the colour of red wine. Depending on the brand, this might be 1 teaspoon (5ml) in a small glass (100ml) of water or more. Put 2.5cm depth of the mixture into each of the small glasses.
2. Trim the roots off the two stalks of celery, leaving each with a nice clean cut at the bottom.
3. Cut the leaves off one of the stalks.
4. Place both stalks simultaneously into the red-dyed water.

5. Now you can choose what to do. You could simply wait – for, say, 5 minutes. Alternatively you could do what I did and encourage the celery with a hairdryer. The idea is to generate a warm breeze over the plant, rather than cooking it, so set the temperature to cool. With the hairdryer you might get a result in less than 5 minutes.
6. When you are ready, take the stalks out of the dye and wash off with clean water.

How to see the results
You have to expose the liquid-carrying vessels just under the skin of the celery.
1. Cut the celery stalk only part way through – about 30mm from the base of the stalk, starting from the inside (i.e. the concave side) of the celery.

You want to get to within 2mm of the outside edge.
2. Starting from your cut, peel back the celery stalk, exposing the inside of the 'skin'.
3. You will be able to see the fluid-carrying vessels – these are the stringy bits in the celery.

How far up the stalk has the red dye gone?

Has it gone further in the plant with, or the one without, leaves?

Did it make a difference if you used a hairdryer?

THE ROOTS
Hales's work with vines showed how the roots help to get sap up trees, but be warned: this only really happens in the spring, and is not easily visible in many plants.

You will need
- A tomato plant in a pot
- Plastic tube with inside diameter a tiny bit smaller than the outside diameter of the tomato plant stem and about 2.5cm long
- Vaseline or other sealant
- A cork, bung or other waterproof stopper to seal the tube, with a hole in it to take a thin (2mm), transparent or translucent plastic straw or other tube
- Stick to support the tube

WHY IT WORKS... Hales realized that the leaves somehow drag water up the plant. It seems that as water evaporates or transpires from the leaves, it pulls up the column of water behind it. It is possible to make an artificial leaf by attaching a porous ball to the top of a tube and blowing air over it to mimic water transpiring from a leaf. It can exert a considerable pull.

What to do

You may find it easier to assemble the tubes before attaching the assembly to the plant.

1. Cut off the plant about 7.5–10cm above the soil with a sharp knife.

2. Push the first tube over the tomato plant stem. It may remove some of the outside layer, but this doesn't matter. It is more important that it seals, and you may want to apply some Vaseline.

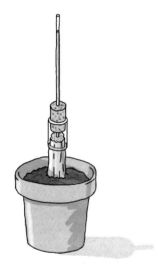

3. Push the straw or small tube through the hole in the bung, and put some Vaseline round the outside to seal. Then push the bung/straw assembly into the larger tube on the plant. As I said earlier, it may in practice be easier to assemble the whole thing and then attach that to the plant.

4. Push the stick into the pot, and tape the straw to it.

5. Keep the pot watered, and see whether you can measure root pressure: i.e. how far does sap go up the tube? Hales used a vine and managed 8 metres.

THE STEM

In fact, you've already experimented with a lone stem using the celery stalks. If you leave the stem without leaves in the food-dye mixture long enough, what happens? Hales found that the stem itself can cause water to move up the plant by capillary action. This only works with tiny tubes, but you can actually see the process at work.

You will need

- 2 small, flat, clean pieces of glass or stiff clear plastic. Microscope slides are great, as are the glasses used to mount your holiday slides for projection. Otherwise, you could try bits taken from the front of audio cassette cases
- Sticky tape
- Food dye mixture as in the leaves demonstration

What to do

1. The pieces of glass must be clean and grease-free. Wash in detergent first, then rinse and allow to dry. You are going to tape the pieces of glass together face-to-face, but held apart slightly at one end.

2. Cut two very small pieces of sticky tape, say 5mm square, and stick to the two corners at one end of one piece of glass.

3. Put the other piece of glass on top. At one end it will rest on the glass underneath, at the other it will rest on the pieces of tape.

4. Fix the pieces of glass together by taping round the end away from your pieces of tape. This should create a small, tapering gap between the sheets.

5. Dip the open end into the dye solution. What happens?

DID YOU KNOW?

Stephen Hales was a bit of a puritan and campaigned for years against drinkers of brandy. He made his sinful parishioners stand outside his church naked – except for a white sheet.

WHY IT WORKS... Capillary action is one of the phenomena caused by the strange things that happen at the surfaces of liquids, and where liquids meet solids. If the liquid is able to 'wet' the solid, this means it adheres and is attracted to the solid to some degree. But particles of liquid also attract each other. In a tiny tube ('capillary' comes from the Latin for hair) or in the very thin gap between glass sheets, the amount of glass surface compared to the liquid surface is very great. If the adhesive force between glass and water molecules is greater than the force between the molecules themselves, the water will be pulled up the gap somewhat. How far up the gap depends upon the balance between the two forces, and weight of liquid being pulled up.

CHAPTER FIVE
COMMUNICATIONS

CARBON PAPER

Making copies of letters and other documents was a tedious task until this idea came along. Like many other inventions, it seems remarkably simple – with hindsight – and the name carbon paper is simple and precise, for it is mainly carbon and paper.

You will need
- 1 teaspoon charcoal (e.g. part of a barbecue briquette, a piece of artist's charcoal, or a dozen burned matches)
- A serving spoon and a teaspoon
- 10ml or 2 teaspoons olive oil or cooking oil
- A small cheap paintbrush
- Some sheets of ordinary (e.g. A4 copying) paper

What to do

WARNING: CARBON IS VERY BLACK, AND THIS CAN BE MESSY!

1. Put a small piece of your charcoal into the big spoon, and use the teaspoon to grind it to a powder by pressing the small bowl into the big one and rocking the small spoon to and fro. If necessary add a bit more charcoal and repeat the process until you have made around a teaspoonful of fine powder in the big spoon.
2. Add 1 teaspoon of oil, and repeat the grinding process to mix the oil and charcoal thoroughly. If necessary, add more oil and repeat the process until the mixture is liquid enough to paint with.
3. Paint the oily black mixture all over one side of a piece of paper, going to within 2cm of each edge.
4. Repeat this with more pieces of paper if you wish.
5. Leave the paper to dry overnight, preferably in a warm place.

Using the carbon paper
1. Lay the paper, painted side down, on a clean sheet of paper, with a third piece on top. Lay the whole sandwich on a smooth hard surface.
2. Write on the top sheet, using a ballpoint pen. Try something simple, such as your name and address. You should find a clear copy on the bottom sheet.

You should be able to use each sheet of carbon paper many times before the copies become too faint to see clearly.

More things to try
Carbon paper is useful for copying letters, but you might like to try some other ideas.
● Make 'brass rubbings'. Put a coin on the table. Lay a sheet of paper on top, and a sheet of carbon paper on top of that. Using the side of a soft pencil, scribble gently all over the surface of the coin. You should find a clear impression or print of the coin on the middle piece of paper. Can you find other objects from which you can make such rubbings?
● Write on a difficult surface such as plastic. See whether you can find surfaces that you are able to mark with carbon paper but not with a ballpoint pen.

WHY IT WORKS... Charcoal is almost pure carbon, and when you grind it up you make particles of pure carbon, like soot, and very black. The oil disperses it and sticks a fine layer of carbon particles to the paper. The process of 'drying' is partly the oil soaking into the paper, leaving the carbon particles on the surface.

When you write on the layer above, you press a line of carbon particles on to the bottom sheet of paper, matching the writing on the top.

FAX MACHINE

Of all the demonstrations we have ever made for *Local Heroes*, this was perhaps the most amazing. The idea that the fax machine was invented not in the 1970s, or even the 1870s, but in the 1840s is odd enough. That it was invented by a Scottish shepherd seems unbelievable. Yet Alexander Bain did patent his 'Electro-Chemical Telegraph' in 1843.

DANGER FROM MAINS ELECTRICITY/ CORROSIVE CHEMICALS

You should tackle this only if you are confident about electrics, and happy to experiment to get the thing working. This is the third generation of our design, and no doubt it can be improved further still. But it does work, and demonstrates beautifully the challenges that modern electronic fax machines still have to meet.

Because this is a complex project, we can't describe every tiny detail. You will in any case need to adapt the design to the materials you have at hand. Rather than dive into the recipe, it will help you to understand how to make the fax if you understand how it works.

Our fax machine looks like a shelving unit because it *is* a shelving unit. The diagram (overleaf) shows the sending machine. The receiver is identical except that the 'picture' is replaced by a sheet of electro-sensitive paper. Information is transmitted from sender to receiver through three circuits: the pendulum circuit, the picture circuit and the escapement circuit.

The pendulum circuit
The weighted pendulum on the sender is held up by an electromagnet. When the electromagnet is released, the pendulum swings. On the receiver, its electromagnet is in the same circuit.

So its pendulum swings at exactly the same time. If both pendulums are the same length, they will be perfectly synchronized. After one swing the pendulum is caught again. Bain used the same mechanism to synchronize the station clocks in Edinburgh and Glasgow.

The picture circuit
Bain's fax used an electrically conducting picture. We made ours from copper printed circuit board. A stylus is attached to the pendulum, and as the pendulum swings it scans across the picture. An electric power-pack is connected to the stylus, so that whenever it passes over the copper bit of the picture, a circuit is completed, but whenever it is over the insulated background, the circuit is broken. The same circuit is connected to the stylus of the receiver, which passes over electrically sensitive paper. When the sending stylus is over copper, the circuit is made and the receiving stylus makes a black mark on the paper.

The escapement circuit
So far, the sender and receiver have scanned back and forth together, the receiver making a black mark wherever the sending picture is copper. But only one line is scanned. So the picture is suspended from an escapement, controlled by a solenoid (an electromagnet with an iron rod in the middle, which is pulled in when the solenoid is activated). When the solenoid 'clicks' it pulls on a lever

that allows the escape wheel to revolve a few degrees, and then stops it again. As the wheel turns, it unwinds the string holding the picture, and it drops a millimetre or so. The next line of the picture is now in position to be scanned. And because the receiving picture is also on an escapement controlled by the same circuit, it will drop an identical amount, at the same time.

The fax is operated by pushing two buttons in turn, the first to release the pendulum and scan, the second to turn the escapement and drop the picture by one line.

You will need
We list the special equipment, and leave bits of wood, screws and bolts up to you.

- 2 wooden shelving units, taller than your pendulum (we have made 1.5m and 2m high versions) – choose ones with adjustable shelves
- 2 lengths of shelf uprights (the sort that takes brackets) as the pendulum – about 1m
- 2 hacksaw blades, to hang the pendulums from
- 2 G-clamps, to clamp the blade in a wooden block
- Lead sheet to wrap round the pendulum as a weight (see note below)
- 2 x 240-volt electromagnets (the sort used to keep fire doors open), for the pendulum
- 2 x 240-volt solenoids (get the most powerful 'pull' you can)

- 6mm threaded rod for the escapement shafts
- A printed circuit board for your picture, about 15cm square (see note below)
- Printed circuit etching kit (see note below)
- Plastic U-section ducting, for the pictures to slide in
- Electrosensitive paper (see note below)
- Thin copper sheet or similar metal sheet (kitchen foil might do) for making a good connection to the electrosensitive paper
- Electrical connector block
- 2 pieces of stiff wire to form the stylus arm (coat-hanger is fine)
- Stiff copper or bronze wire for the styluses themselves
- A 240-volt push-to-break switch, to release the pendulum electromagnet
- A 240-volt push-to-make switch, to activate the solenoid
- An electrical 'power pack' – the sort used in school labs (see note below)

Notes

Any compact weight should do for the pendulum – the sheet is good because you can adjust it to the electromagnet's pull.

Using the printed circuit board – the resolution isn't very high, so tiny pictures won't show up very well.

The etching kit marks the bits you want to stay copper, and dissolves the background away. You may be able to get an electronics buff to do this for you.

WARNING: THE CHEMICALS ARE CORROSIVE.

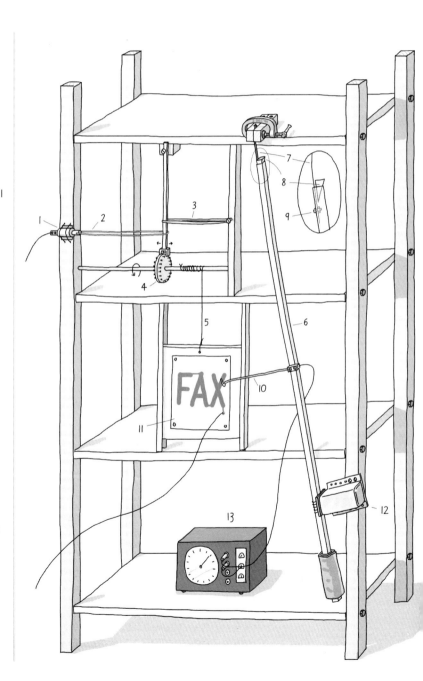

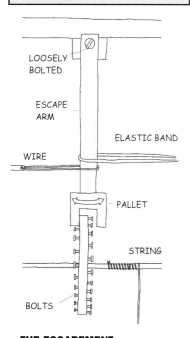

LOOSELY BOLTED

ESCAPE ARM

WIRE

ELASTIC BAND

PALLET

STRING

BOLTS

THE ESCAPEMENT

The electrosensitive paper is the only really specialist item: it's metal-coated paper, and when contact is made it 'burns' away to leave a black mark. You can get various grades, with different required voltages. We chose 20 volt rather than 200 volt.

The exact specification of the power pack will depend upon the electrosensitive paper you use – we needed 30 volts at 5 amps.

What to do

Assemble the shelving units.

The pendulum

1. This is a shelf upright, suspended from a hacksaw blade. Grip the hacksaw blade between two blocks of wood, and screw or bolt the blocks together firmly. Hacksaw blades usually have little 'lugs' at the ends to stop them slipping. These should of course be above the blocks.
2. Now fix the hacksaw blade to the pendulum shaft. If your blade has a hole in it, so much the better. If not, you will have to drill one. Two would be best. Take care – the blade is hard, slippery and sharp. You might consider clamping it between two bits of wood first. Bolt the blade to the shelf upright through the holes in the upright.
3. Now fix the wooden blocks to the top shelf. Clamp first, to make sure you get enough clearance when you have fixed the weight to the pendulum.
4. If you're using lead sheet as a weight, wrap a kilo of it round the bottom of the pendulum. Make sure it won't slip – use a bolt if you like. Swing the pendulum, making sure the weight doesn't bash into anything.

You could fix the support block now, or leave it until everything else is in place. Moving the pendulum in and out is one way of adjusting stylus pressure.

The electromagnet

1. This is fixed to the frame of the shelf unit, its 'keeper' or steel plate attached to the pendulum. Fix the magnet as close to the bottom of the pendulum as you can, to reduce the leverage.
2. Bolt the keeper to the pendulum shaft at the appropriate point, then fix the electromagnet to a metal bracket and fix this to the frame. Your particular magnet may fix differently – perhaps directly to the frame. Make sure you can get it at the right angle, so that the keeper is flat on to the magnet.

The escapement

1. This can be a bit tricky. We made the escape wheel of 4mm ply, 100mm diameter. Around its perimeter you attach little bolts, which project from the surface by about 6mm. The bolts alternate, sticking out first one side, then the other. You should countersink the nuts so that they don't protrude. You will need about 20 bolts, evenly spaced round the circumference, so 10 will poke out one side, 10 the other. When the escapement is clicked on, the arm is pulled and the wheel releases, only to be caught on the next bolt on the opposite side. Then as the solenoid returns, it releases to be caught by the next bolt on the first side. So for each action of the solenoid, the wheel turns a tenth of a revolution.

2. Firmly fix the escape wheel to the threaded rod, and mount in two drilled holes in the main upright and in a secondary upright installed for the purpose (see main diagram).

3. Make the escape arm from some, say, 15 x 15mm timber, and loosely bolt to the top shelf, directly above the escape wheel. On the end of the arm, you need a 'pallet'. This is a square U-shape, attached to the end of the arm, so that when it is pulled from side to side it will engage with the bolts on just one side of the escape wheel, leaving the other side free. So if the bolts protrude 6mm and the wheel is 4mm thick, the gap between the arms of the 'pallet' should be over 10mm but less than 16 mm. Screw the pallet to the end of the escape arm.

4. Fix the solenoid to the main upright of the shelf unit, directly above the level of the pallet. The exact position may vary depending upon how far the solenoid pulls. Wire the solenoid shaft to the escape arm. This will pull it to the left. Fix an elastic band to pull it back to the right.

The picture

1. Design something nice and bold, and then make a printed circuit version of it. You can buy complete printed circuit kits, which certainly make life easier. The important thing is that all the parts of your picture are connected together. There must be no 'islands' of unconnected copper. This may make the design look a bit odd, but at least it will all transmit properly. Leave one 'blob' as a connector, and solder a wire to it.

2. The printed circuit is mounted on a piece of MDF or chipboard, which runs in smooth plastic U-shaped channels. We used plastic conduit. The channels are fixed to extra wooden uprights, which are in turn fixed to the shelf unit. The centre of the picture should hang below the centre of the right-hand portion of the threaded rod that acts as the axle for the escapement.

3. The picture is hung from string wrapped round the rod. Tie the string with the picture in its lowest position, then wind it up by turning the escape wheel. You'll have to jiggle the escape arm as you do this.

The receiving picture

1. Tape the electrosensitive paper to a piece of MDF like that used to mount the printed circuit. You will also need to clamp a large copper (or other conducting) metal connector to the metal surface of the paper, in order to make a circuit. Use a piece, say, 25mm by the full width of the picture, clamped firmly under a piece of wood bolted to the MDF. Wind the bared end of a connecting wire round one of the bolts so that it is firmly clamped to the copper.

The stylus

1. Friction in the stylus is the one thing likely to upset the smooth working of your fax – and yet you need good contact between stylus and copper or paper in order to get enough current. Fiddling and adjusting is the only way.

2. Fix a bit of electrical connector block to the side of the pendulum where you want the stylus to be. When the picture is 'up', the stylus should scan the lowest line of picture. Into the left side of the block, connect a bit of bicycle spoke or coat-hanger – a stiff, conducting wire. On to the end, solder a bit of stiff copper wire – the stylus itself. It will bend in towards the picture, but then have a curve on the end to ensure smooth contact. It has to move smoothly in both directions.

3. Into the other side of the block, connect the wire from the power pack. This wire could impede the pendulum, so you could lead it right up the shaft and take the connection off at the top.

Wiring

When you have made two identical fax machines, you are ready to wire them up. Take care: mains voltages are involved, and even the picture circuit can give you a nasty shock.

The pendulum circuit

1. See the wiring diagrams. Mount the push-to-break switch near to the pendulum, on the upright. Connect the live side of the mains supply (from a fused plug) to one side of the push-to-break switch. From the other (switched) side, take the supply for the two electromagnets – one on the sender, then on to the other on the receiver. Connect the neutral side of the supply to the remaining terminals of the electromagnets, i.e. first to the sender electromagnet, then from that to the receiver.

MAKE SURE ALL TERMINALS ARE INSULATED AND CANNOT BE TOUCHED.

2. Test by pulling the pendulums till they stick to the electromagnet. Push the switch briefly. The pendulums should swing, and then stick again. If they don't stick (and you had

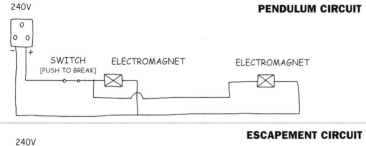

240V

PENDULUM CIRCUIT

SWITCH [PUSH TO BREAK]   ELECTROMAGNET   ELECTROMAGNET

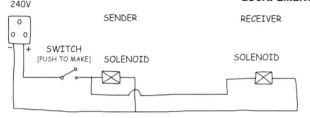

ESCAPEMENT CIRCUIT

240V

SENDER   RECEIVER

SWITCH [PUSH TO MAKE]   SOLENOID   SOLENOID

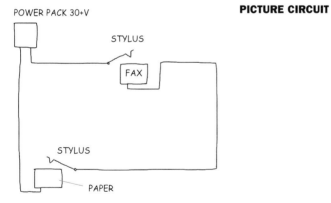

PICTURE CIRCUIT

POWER PACK 30+V

STYLUS

FAX

STYLUS

PAPER

The picture circuit

1. Take the positive output of the power pack, adjusted to suit your paper – we used 30 volts. Connect to the stylus of the sender. Take the wire from the picture on the sender, and connect to the stylus on the receiver. Take the wire from the picture (electrosensitive paper) on the receiver, and connect to the negative side of the power pack, completing the circuit.

Operation

1. Make sure all the electrical connections are secure, and connected to properly fused plugs. Wind the sending and receiving pictures fully up. Turn on the power. Pull the pendulums to the side. Stand by the sender.

2. Push the pendulum button (push-to-break switch) briefly. The pendulums will release simultaneously. As the sending stylus touches copper, you should see a black mark on the receiver's paper. The pendulum returns and holds.

3. Push the escapement button (push-to-make) switch, and then release again. The escapement advances with a double kerr-lunk. The picture drops by 2mm. Start again.

Finally

You probably won't get this to work first time because getting the balance right between friction and good contact in the stylus is not easy. But if you do succeed, you will have made one of the most satisfying of *Local Heroes* demonstrations – and one of the most surprising inventions ever.

released the switch) you will have to adjust the position of the electromagnet, keeper, or pendulum.

The escapement circuit

1. This is rather like the pendulum circuit, but instead uses the solenoids and the push-to-make switch. Mount the switch near the escapement.
MAKE SURE ALL TERMINALS ARE

INSULATED AND CANNOT BE TOUCHED.

2. When you push the switch in, the solenoid will pull the escape arm to one side. The wheel will click round one bolt. Release the switch and the elastic band will pull the escape arm back again. The wheel will click round another bolt. The wheel is pulled by the weight of the picture, pulling on the string.

MEGAPHONES

In the 17th century Samuel Morland investigated long-distance communication, using what he grandly called his tuba stentorophonica and what we call a megaphone.

You will need
- A newspaper – or, slightly better, 2 square metres of flexible plastic sheet
- Tape
- A friend, assistant, or accomplice

What to do

1. Roll up your plastic sheet – or two sheets of newspaper – to make a long cone. The narrow end of the cone should have a diameter of about 7–8cm, to fit comfortably round your mouth when it is wide open. The wide end should have a diameter about 15–20cm. The cone should be as long as possible, without being too floppy.

2. When you have the shape about right, tape the edges roughly so that the cone does not unroll again at once.

3. If your assistant is keen, make a second cone the same as the first. To try out your tubae stentorophonicae you and your friend should have one each and stand a long way apart – say, 50 metres at least. The human voice is hard to hear over long distances, although the audibility varies a lot according to whether you have buildings, mountains, water, traffic and other noise, and even with the air temperature. The best place to find is some open ground, perhaps a park or an empty playing field – but you can try this on the beach, across a river, or even along a footpath.

4. Put the narrow end of the megaphone to your mouth, point it where you want the sound to go, and shout clearly, but do not scream. Make sure you hit all those p, t and d sounds. Leave a gap after each sound.

5. You may simply be able to hold a conversation, but for a more scientific test you might like to set up a specific plan. One of you might like to shout three numbers between 1 and 10 – say, 7, 3 and 4. The other one writes them down, shouts three numbers back, and then you can meet and compare notes. If you got them all right, move further apart and try again.

6. One neat way of getting instant feedback is to shout the signals code-words for letters of the alphabet, and get your assistant to signal each one back using semaphore. You know the kind of thing – Alpha, Bravo, Charlie, Delta, Echo, Foxtrot, Golf and so on. You shout, 'Charlie', and if your assistant then makes the semaphore 'C', you know you must be audible.

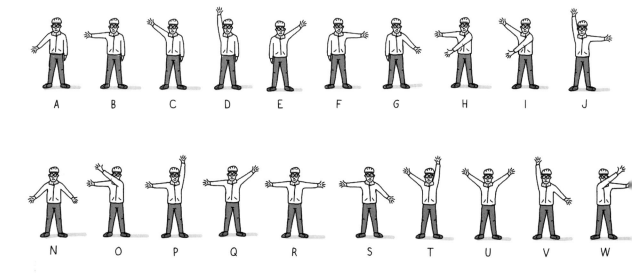

More things to try

● How far can you make yourself heard with the letters of the alphabet?
● At what distance can you hold a conversation?
● What kinds of people have the most audible voices? Children, women or men?
● Try making one megaphone much longer than the other. Does that make any difference?
● What about the width of the wide end?
● Can you make megaphones from different materials – or spray the inside of one with lacquer? What difference does it make?
● When might megaphones be useful to you?
● Can voices really be heard for miles over still water at night?

DID YOU KNOW?

Samuel Morland was a bit of a pompous twit, but he did save the life of Charles II. Working as a spy in 1658, he heard Cromwell discussing a plot to have the king murdered. Morland warned the king, who was, naturally, most grateful. Eventually he knighted Morland and gave him the title of Magister Mechanicorum (or Master of Mechanics). As well as making serious scientific investigations, Morland invented some ludicrous things – such as a calculating machine described by the physicist Robert Hooke as very silly, and a portable clockwork kitchen!

WHY IT WORKS... When you shout without a megaphone there is poor coupling of the vibrating air in your mouth and all the air outside. So much of the effort is wasted, and much of the sound goes off in the wrong directions.

Using a long megaphone that fits your mouth effectively turns it into a long tube, and the vibrating air in your mouth is well coupled to the air in the tube, so that all the air in the tube vibrates. Much less effort is wasted.

What's more, the big end of the megaphone now acts as a mouth, and projects the sound accurately in the direction you want. So the sound that leaves the megaphone is much more energetic and directional than any shout you can make without it.

Morland claimed that using a megaphone 20 feet (6 metres) long he was able to hold conversations at a distance of almost a mile. Using a megaphone say half a metre long, you should be able to make yourself heard at 100 metres.

MECHANICAL TELEGRAPH

How do you send messages over great distances, without using electricity? During the Napoleonic wars – up to 1815 when Wellington won the Battle of Waterloo – commanders needed to get orders to their troops. Lord George Murray devised a telegraph that could send messages from London to Portsmouth in just one minute.

The size of telegraph you make affects the range: the bigger the flaps, the further away they can be seen. This design assumes 30cm-square flaps – more than big enough for signalling over 100 metres.

You will need

- 6 pieces of 25 x 25mm timber, 30cm long
- Screws
- Tacks or staples and glue
- For the flaps: 6 x 28cm squares of hardboard, thin ply or corrugated plastic stiffener, painted black
- For the uprights: 3 pieces of 50 x 25mm timber, 1.5m long
- 10 pieces of 50 x 25mm timber, 30cm long
- Fixing brackets (optional – you should get away with nails and glue)
- 6 x 2m lengths of string
- 6 round fishing weights or small blocks of wood to act as counterweights

What to do

1. First prepare the flaps. Into the centre of each end of the six pieces of 25 x 25mm by 30cm timber, drill a pilot hole and drive in a woodscrew, leaving about 1cm protruding. This forms one part of the bearing, on which the flaps will rotate.
2. Now fix the flaps centrally on these pieces of wood, using tacks or staples and glue.
3. Next, mark up the uprights. Lay them out side by side on the ground,

the 50mm side up. Mark across all three uprights where the cross-pieces of the frame will go. Remember, you need to leave 30cm-square holes. So the first one is at the top, the next comes 30cm below it, and so on until you have the position of all the cross-pieces marked. Now turn the middle upright over, and mark the positions again on the other side.
4. Next, make the holes for the flaps to rotate in. These will be half-way between the cross-piece positions you have just marked. The holes have to be wide enough to take the heads of

the screws you put into the flap supports, and for them to rotate without sticking. You should again mark a line across all three uprights where the holes are to go, and then you can set up a nice production line. DO NOT drill all the way through the wood. You are making a little cup for the screw heads to sit in: 5mm depth should be adequate.
5. Remember to turn the central upright over and to drill its three holes on the other side too. It's not a disaster if the holes in the central upright go right through.

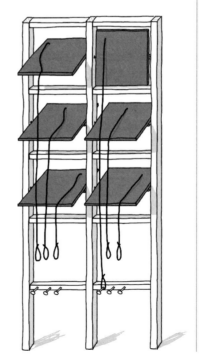

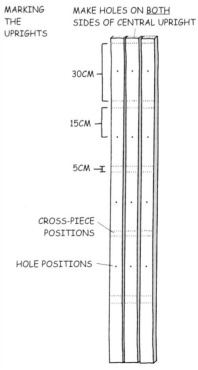

MARKING THE UPRIGHTS

MAKE HOLES ON <u>BOTH</u> SIDES OF CENTRAL UPRIGHT

30CM

15CM

5CM

CROSS-PIECE POSITIONS

HOLE POSITIONS

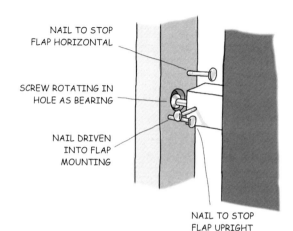

NAIL TO STOP
FLAP HORIZONTAL

SCREW ROTATING IN
HOLE AS BEARING

NAIL DRIVEN
INTO FLAP
MOUNTING

NAIL TO STOP
FLAP UPRIGHT

6. Now assemble the frame. Starting with the right-hand upright, lay it on the ground, together with the cross-pieces at their marked positions. Screw from the outside of the frame into the cross-pieces, adding wood glue before tightening.

7. Put the three right-hand flaps into position, attached wood supports downwards, screws engaged in the holes of the upright. Offer up the central upright, getting all the screws into their bearing holes. Holding or clamping the upright in position, test that the flaps will rotate freely – but not fall out. If necessary, adjust the depth of the screws. Finally, screw and glue the central upright in place.

8. Now add the left-hand set of cross-pieces. You can't screw these from the right-hand side because the cross-piece is already in position, unless you fancy some joinery. So either fix these with brackets, or drive in nails from the front of the upright, at an angle into the cross-piece. Put

PVA wood glue in the joint.

9. Put the remaining flaps in place, and offer up the left-hand upright. Check screw depth as before, then screw and glue the final upright in position.

10. Next, rig the strings, fix the weights, put in the stops. Make a small hole centrally in the top edge of each flap. Then just below this on the same side as the flap support, fix the counterweight. This will make sure the flap tips into the horizontal position when at rest.

11. The stops are nails that stop the flap at the fully horizontal and vertical positions. Drive one nail into the flap support so that the nail is horizontal when the flap is vertical. Now drive two nails into the upright to engage with the nail, as shown in the diagram.

12. Tie lengths of string through the holes in the upper edge of the flaps, and lead them down to the lower cross-pieces. If you pull the strings, the flaps move to the vertical (closed)

position. Make a loop in the string where it meets the cross-piece, and put in a nail for the string to hook over. Repeat for all six flaps, then trim the strings below the loops.

Mounting the telegraph

This depends where you intend to use it. It could be propped in a window, mounted in a tree (with extended strings) or fixed to fence posts.

Time to signal

At last you can send messages. All you need is the code. As you can see overleaf, the telegraph uses an alphabetic code. This is slightly surprising, since anyone could have stood next to one of the stations and intercepted the messages. However, the thinking was that because the messages travelled with such incredible speed, by the time a spy had got the contents to Napoleon in France, several days would have passed and it wouldn't much matter.

Here is the sending procedure. The resting position is 'all white'. When you are about to send, you put up 'all black'. If there are two stations – or a string of them as the navy used – the next station repeats the signal just sent. Then you know the message is getting through. So station one waits until it sees each letter or a special signal repeated by station two before sending the next one. The alphabet lacks a J or a U, as in Latin. Use I or V respectively.

Actually, there are 63 combinations: the six flaps can each have two positions, so think of it as binary system – 1, 2, 4, 8, 16, 32. The next (seventh) position would be 64, which means that with six flaps you can have all the numbers up to 64, ie 63. The admiralty took advantage of this extra capacity by devising special signals. You may not have much use for the 'Court martial to sit and try offenders' or 'Put sentence of court into execution' codes. But you could substitute 'Your dinner's ready' or 'Milk and two sugars, please' or 'No, really, I've done all my homework'.

How the message got through

The telegraph stations were set in a line between London and Portsmouth, between eight and twelve miles apart. The fastest message was the Greenwich Time Signal, which took just a minute to get to Portsmouth. Some messages could take 25 minutes, and had to be 'held' at intermediate stations if the weather became too foggy.

Fishy business

At one time the Admiralty panicked when they thought the system had been infiltrated. In fact it turned out that the operators had their own private code and the sinister secret message was PRAWNS.

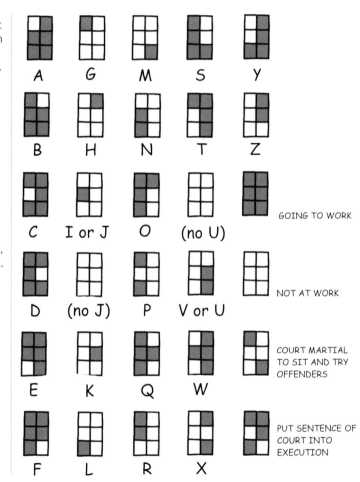

A G M S Y

B H N T Z

C I or J O (no U) GOING TO WORK

D (no J) P V or U NOT AT WORK

E K Q W COURT MARTIAL TO SIT AND TRY OFFENDERS

F L R X PUT SENTENCE OF COURT INTO EXECUTION

WATT'S COPYING MACHINE

While he was working in partnership with the entrepreneur Matthew Boulton, James Watt designed many steam engines and other devices. He got so bored with having to copy all his design drawings that he invented a copying machine to speed up his work – and you can make your own version (in case the photocopier goes on the blink!).

RATING: ✳ EASY!

What you need
- Half a teaspoon of sugar (icing sugar is good but any other will do)
- A small container, e.g. a bottle-top or egg-cup
- Black ink
- A pen you can dip into the ink – a quill pen will do! (See below for how to make your own.)
- Ordinary (copying) paper
- Tracing paper (you may find that greaseproof paper or baking parchment will work – or try ordinary paper)
- A thin squidgy mat – a mouse mat works well, or try a folded newspaper
- A rolling pin or bottle (a milk bottle or wine bottle will do, but it must have straight vertical sides)

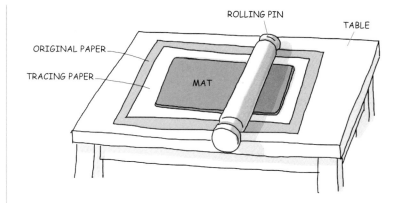

What to do
1. Put the sugar into the bottle-top or egg-cup and add about a teaspoonful of ink. Stir until the sugar has dissolved and the ink is slightly thick. If the ink is too thick to write with, add more fresh ink.

2. Write a message and/or make a drawing on a piece of paper, using your dip pen and the sugary ink. Leave it to dry for a few minutes, until the ink doesn't look wet.

3. While the ink is drying, make a piece of tracing paper damp. The best way to do this is to spray it with a fine mist of water (e.g. from a small plant spray), but you could also sprinkle water on it gently and then wipe it with a damp cloth. Your tracing paper should be just damp, but not soggy.

4. Lay the original message on a smooth hard surface, face up, lay the damp tracing paper on top, and lay the squidgy mat on top of that.

5. Using your rolling pin or bottle, and pressing down firmly, roll across the mat and back. One pass in each direction should be enough, if your two sheets of paper are well prepared.

6. Take off the mat, and carefully peel the tracing paper away from the original. You should have a clear copy on the tracing paper. This will be reversed left-to-right, so you will have to read it from the back.

More things to try
● See how many copies you can get by trying several pieces of tracing paper, one after the other.
● Can you use two pieces of tracing paper, one on top of the other?
● Do you get a better copy by rolling to and fro several times?
● Can you use ink of any other colour?

How to make a quill pen (in case you need it!)
1. Find a big strong wing feather, ideally from a goose or a seagull, but a chicken feather will do.
2. Cut off the very tip of the shaft at an acute angle of 30–45 degrees. For an italic nib, cut the tip across to make a flat tip about 2–3mm wide.
3. Make a cut from the tip about 5mm up the shaft.

DID YOU KNOW?
This is exactly the process that James Watt devised and patented in 1789. Matthew Boulton was enthusiastic about the idea, and the company probably made good money from it. What was more important was that they revolutionized office practice, because everyone could now keep copies of vital documents. A useful benefit was that we know a great deal more about the Boulton and Watt business than about earlier companies, because they kept copies of all their business documents.

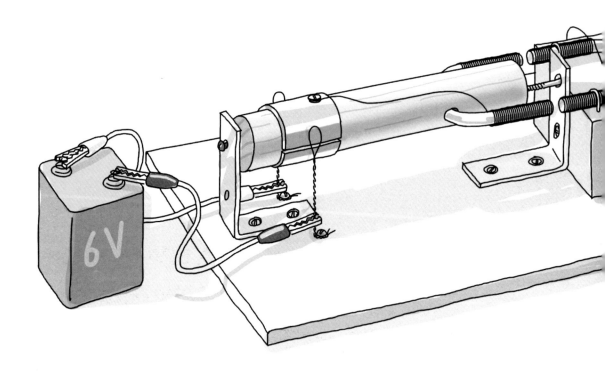

CHAPTER SIX
MECHANICS

AUTOMATIC EGG BOILER

Of all the inventions we have featured in *Local Heroes*, none is more archetypal than the Automatic Egg Boiler patented by Dr Thomas Gaddes of Harrogate in around 1896.

RATING: ✳✳✳ CHALLENGING

DANGER: HOT WATER/ HEAT SOURCE

The patent this design is based on is for 'modifications to automatic egg-boilers'. Dr Gaddes had taken out the original patent the year before, and although it would certainly have worked, it was much more complicated. Dr Gaddes must have seen the light some time in 1896–7 and realized that he could do away with sliding pivots and various other refinements, in favour of this simpler and more elegant design.

This is a great demonstration to do in public. As water drips slowly from the reservoir, tension mounts in the audience. The arm tilts – then nothing happens. Has it broken? Will it work? Then, the laws of physics take over: the weight slides and with a clatter the egg lifts, invariably to a huge round of applause.

You will need
- Saucepan
- Screws
- Wood for upright and pivoted arm, approx 25mm x 18mm, one 30cm long, the other 40cm
- Plastic shampoo bottle, 500ml size (empty)
- Stiff wire, e.g. coat-hanger
- Bathroom plug chain, or soft string
- Fishing weight (approx 50g)

WARNING!
TAKE CARE TO STABILIZE THE EGG-BOILER. THERE IS A REAL DANGER OF IT OVER-BALANCING AND TIPPING SCALDING WATER OVER YOU, OR EVEN SETTING YOU ALIGHT IF YOU ARE USING A CAMPING STOVE.

What to do
1. Choose a saucepan with a wooden or flat plastic handle. Drill two holes in the handle fairly close to the pan, and use screws to fix the upright to the handle. We made ours about 30cm tall. The seesaw is another piece of wood, somewhat longer. Drill a pivot hole in it so that when pivoted from the upright, one end will be over the centre of the saucepan: this is where you want to hang the egg. Drill the seesaw where it is to pivot. Fasten to the top of the upright with a bolt, or a screw slightly smaller than the hole in the seesaw, and not screwed fully down, so that it is loose enough to pivot. The screw should, however, fix tightly to the upright.

2. Next, construct the reservoir. Cut off the bottom one-third of the bottle. Drill a hole (about ³⁄₁₆ of an inch according to the patent – say 5mm) in the bottom part of the bottle, at the edge. This is your reservoir. Use a piece of coat-hanger to suspend the reservoir from the end of the see-saw, drilling small holes in the see-saw and reservoir for the wire to pass through. The wire should be

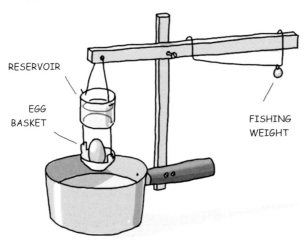

RESERVOIR

EGG BASKET

FISHING WEIGHT

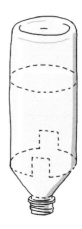

V-shaped, the point of the V being at the seesaw. Fasten by bending up the ends of the wire after they pass through the reservoir.

3. Make the egg-basket by cutting off the neck of the bottle, and making two suspension points by cutting the bottle as shown.

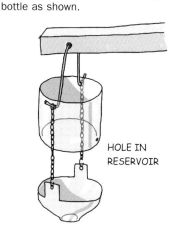

HOLE IN
RESERVOIR

4. Use a piece of wire to make an end-stop for the seesaw. You don't want the egg-and-reservoir to pull it more than 15 degrees below the horizontal. Drill a hole in the upright and stick in a piece of wire as the end stop. Attach the egg-basket to the reservoir, using pieces of bathroom chain. Make sure the chain is long enough to completely submerge the egg when the see-saw is on its end stop, but short enough

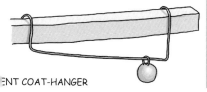

ENT COAT-HANGER

that the egg is right out of the water when the arm tips the other way. The idea is that the egg sits on the bottom of the saucepan when ready for cooking, but the reservoir is always suspended above the surface of the water.

5. Finally, make the weight assembly out of a coat-hanger. It should have a hook at each end, and the long side should be at about 15 degrees to the seesaw. Slide the fishing weight on to the wire, then attach it to the seesaw by hooking it over at either end. Fix with tape until you have determined the right position, when you can make it more permanent with fixing grooves in the seesaw for hard, medium or soft eggs.

6. Now you are ready for testing. Fill the saucepan with water and heat to a simmer. Fill the reservoir with water, and load an egg. Make sure the weight is at the saucepan end of its wire. The water and egg should pull the arm down until the egg is under water and the chain is slack. As water drains from the reservoir, the arm will eventually begin to slowly rise. The chain is long enough to leave the egg undisturbed. However, as the weight assembly tips below the horizontal, the weight will suddenly slide along the wire, and pull the egg out of the water. In theory you should be able to calibrate the system for 4 minutes in a central position, then slide the weight-and-wire assembly up or down for hard or soft eggs. Make sure the chain is long enough for the wire to dip below the horizontal, and that the weight slides smoothly.

A HARD-BOILED DENTIST

I like to imagine the very busy Dr Gaddes in his Harrogate dentist's surgery, trying to make his breakfast but constantly called away by patients with toothache. He must have been so frustrated by the diet of hard-boiled eggs that he felt driven to design this ingenious device. It is clear from the patent that he really did make it – the design has all sorts of little features you wouldn't know you needed until you tried. For instance, he describes in great detail a cunning little gutter attached to the reservoir outlet to divert the flow towards the edge of the saucepan – so that it didn't cool your perfectly cooked egg!

BICYCLE CLOCK

The Dutch scientist Christiaan Huygens grabbed Galileo's clever ideas about the pendulum (see page 126) and built the first pendulum clock in 1645, just after Galileo died. To celebrate their genius we turned a bicycle into a clock.

You will need

- A bicycle with a rack over the back wheel
- A stand to hold the back wheel off the ground and stop the bike falling over (see below)
- Approx 2m softwood, say 20 x 30mm or 20 x 40mm (or the same length of Dexion)
- Nuts, bolts, screws, clamps
- 2 weights, one approx 1kg and the other less – e.g. a half brick
- 16cm length of thin aluminium strip, 25mm wide (for the pallets)
- String or wire

For the stand, you could use the kind that bike mechanics use, or make your own with 8 pieces of Dexion 10–40cm long, or 3 pieces of softwood approx 40 x 15 x 2cm, and some short offcuts of softwood, say 2 x 3cm.

What to do

1. To make a simple stand: either use the Dexion, and bolt eight pieces together like this:

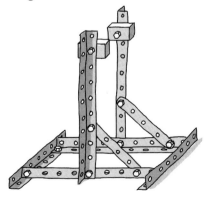

2. Or use wood, and screw three pieces together like this:

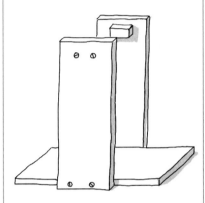

Fix a 5cm length of softwood offcut near the top of each upright to form rests for the frame just in front of the back axle. Make sure that the uprights are just about as wide apart as the length of your back axle (about 17cm), and the supports are at the right height (about 35cm for a mountain bike) to hold the back wheel a few centimetres off the ground.

The pendulum

1. Make the pendulum from a piece of wood or Dexion about 1m long. At the top suspend it from a screw or bolt so that it can swing freely at right angles to the back wheel. From the bike rack build up a simple structure in order to suspend the pendulum 60cm above the rack and 10cm behind the back wheel, with the end of the pendulum at the same height as the back axle. Use bolts or clamps to fix the pendulum support to the rack.
2. Across the bottom of the pendulum fix a 30cm bar, and to the ends of

this fix 10cm spurs, facing forward.
3. On these spurs fix the pallets (this is a technical term used in clock-making – nothing to do with builders and fork-lift trucks!). We made ours from the thin aluminium strip, with the ends crudely shaped to a blunt

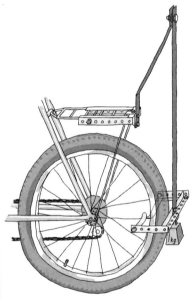

point and bent up at about 30 degrees. These ends should overlap by about 1cm, and one pallet tip should be about 1cm above the other, but keep the pallet-fixing flexible because you will have to adjust the pallets when the whole clock is set up.
4. Use wire or string to fix the weight to the very bottom of the pendulum.
5. Fix the smaller weight to one of the pedals with tape or string, so that it pushes the pedal round.
6. Put the bike in top gear. Adjust the pallets so that they go through the

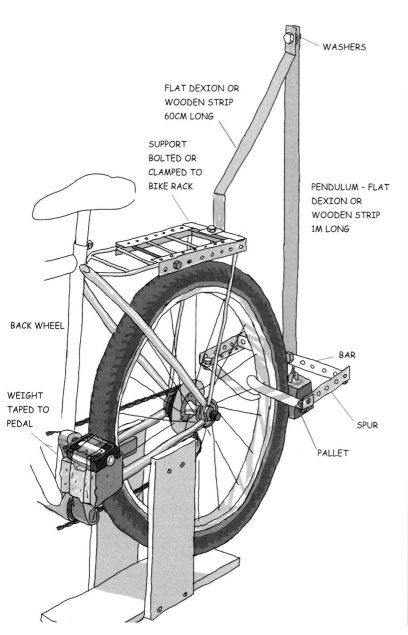

WASHERS

FLAT DEXION OR
WOODEN STRIP
60CM LONG

SUPPORT
BOLTED OR
CLAMPED TO
BIKE RACK

PENDULUM – FLAT
DEXION OR
WOODEN STRIP
1M LONG

BACK WHEEL

BAR

WEIGHT
TAPED TO
PEDAL

SPUR

PALLET

spokes of the back wheel and overlap by about 1cm.

7. Lift the weighted pedal to the top of its travel, start it off and swing the pendulum. The pendulum should swing to and fro, allowing one spoke to escape through the pallets on each swing. This is your 'escapement mechanism'.

8. You will probably have to adjust the positions and the bend of your pallets to get it to work, and they may jam on the valve, but with care you should be able to get your clock to run for a minute or so, until the weighted pedal reaches the bottom.

9. Notice that the tips of the pallets must be bent upwards, so that each time a spoke hits one of them it gives the pendulum a push. This is how the power source keeps the clock going.

You have now put together the three essential components of the Huygens pendulum clock: a regular timekeeper (the pendulum), a power source (the weight on the pedal) and a regulator to couple the two and count the swings (the back wheel). Put a sticky label on the tyre by your valve, and you can use the back wheel as the hand of your clock; it should make one complete revolution in a fairly precise length of time.

In practice the mechanism is crude, and you should not expect to have a precise clock.

More things to try

What difference does it make if you:
● Vary the weights?
● Put the bike in a different gear?
● Vary the length of the pendulum?

FARADAY ELECTRIC MOTOR

Michael Faraday, one of the greatest scientists of all time, invented the world's first electric motor in September 1821. You can make a motor just like his.

You will need
- A piece of wood or MDF about 15–20cm square
- 10cm of softwood, 20mm x 30mm
- 1cm dowel, 20cm long
- A metal dish about 15cm across. We used an old meat-pie tin, but you could probably use a large baked bean tin, a bare metal roasting tin, or a metal foil take-away container
- A brass eye or hook
- Bare copper wire, say 2mm and 20cm long – we salvaged this from an old piece of power cable
- A powerful bar magnet – it does need to be powerful, and should pick up at least six paperclips
- Insulated copper wire – about 50cm
- Big batteries to deliver at least 12 volts, and preferably 24 volts
- Crocodile clips or bulldog paper clips
- Salt – say 100g
- Warm water – hot water from the tap is fine

What to do
Make the stand
1. Drill a 10mm hole in the corner of your square piece of wood to fit the dowel. Drill another near the end of the piece of softwood.
2. Glue and/or pin the dowel into these holes so that the metal dish can stand on the wooden base and the softwood hangs over the middle of the dish.

If you have no dowel you can use ordinary softwood for the upright, and fix it with nails or screws.

Set up the circuit
1. Screw the brass eye or hook into the underside of the softwood, so that it is above the middle of the base.
2. Place the metal dish on the base.
3. Straighten the bare copper wire as well as possible, then bend one end into a smooth round hook. Hang this on the brass eye. Trim the straight end of the wire until the bottom end hangs just not touching the bottom of the metal dish. The wire should be free to swing in any direction.
4. Stand the magnet on end in the middle of the tin, where the wire is hanging.
5. Using your insulated wire, connect your batteries together in series, positive to negative, in a row.
6. Connect one end to the brass eye, without fouling the hanging wire; a

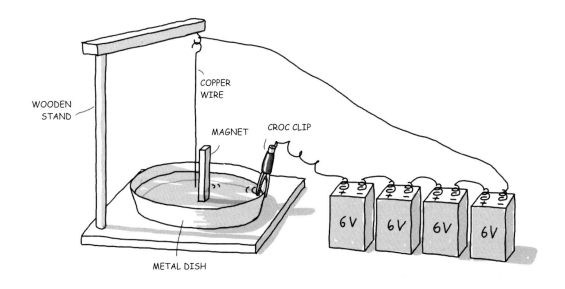

COPPER WIRE

WOODEN STAND

MAGNET   CROC CLIP

6V   6V   6V   6V

METAL DISH

crocodile clip may be useful, but it is not essential.

7. Connect the other end to a crocodile clip or bulldog clip. Do not complete the circuit yet.

8. Stir the salt into the water until most of it dissolves.

9. Pour the salty water into the tin. The water must be deep enough for the end of the copper wire to hang in it.

10. Now you are ready to go. To complete the circuit, clamp the crocodile (or bulldog) clip to the side of the metal dish, preferably below the water level.

The bare copper wire should swing in a circle, round and round the magnet, and you should see bubbles appearing in the water. These bubbles are the gases oxygen and hydrogen, made by electrolysis of the water.

More things to try

● What happens when you turn the magnet upside-down? Can you explain what you see?

● What difference do you think there will be if you use a stronger battery, a stronger magnet, or a stronger salt solution?

● Try using pure (not salty) water. How many drops of lemon juice do you need to add for the motor to work?

WHY IT WORKS... When a current flows in a wire (like your bare copper wire) it makes a magnetic field in a cylinder, like a sleeve, around itself. This magnetic field interacts with the field from the bar magnet to push the wire round in a circle.

This Faraday motor is only a toy - the first useful motor is on the next page.

STURGEON ELECTRIC MOTOR

William Sturgeon (1783–1850) invented the first practical electric motor, and now you can try making your own version.

You will need

- 2 x 6cm steel U-bolts (from a hardware shop)
- 20cm of wooden broom handle (30mm diameter)
- 4cm of 30mm copper pipe for the commutator (a connecting piece will do)
- Screws
- 500g roll of 1mm enamelled copper 'winding' wire
- Wooden mounting block (a cube slightly smaller than your brackets – see below)
- Wooden baseboard, approx 40 x 30cm (hardboard will do)
- Wire staples or wire, for fixing
- Small (10cm) mounting brackets (L-shaped)
- 60cm of springy wire (ideally bronze picture-hanging wire)
- 1m of insulated electrical wire
- 8 crocodile clips (optional)
- 2 x 6-volt lantern batteries (996 size, the ones with springs on top) – alkaline are best

What to do

1. Start by fixing one U-bolt to the broom handle. This is quite tricky. You need to drill a hole through the shaft of the broom handle, so that you can push the U-bolt through it. Drill at an angle from both sides, but make sure the hole is not too big. Ideally the bolt

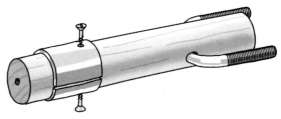

will push through with difficulty, then sit fairly rigidly. The ends of the bolt should project a clear 1cm beyond the end of the shaft. So, with a 6cm bolt, the hole should be 5cm from the end of the shaft.

2. While you've got the drill out, carefully drill a small (3–4mm) hole exactly in the centre of each end of the handle. If the hole is off centre, then the motor will wobble all over the place!

3. Now fix the copper pipe to the broom handle. If the pipe doesn't slide over the end of the broom handle (away from the U-bolt), you'll need to whittle the end of the handle slightly until the pipe fits snugly on. Push it about 1cm along the handle. Fix it in place with two small screws opposite each other. Important: the screws must be placed perpendicular to the plane of the U-bolt.

4. Using a hacksaw, cut right through the pipe on both sides. Important: the cuts should be in line with the arms of the U-bolt. The two screws should now hold the two halves of the pipe on the shaft. The cut pipe is the commutator – the crucial component that will alternate the direction of the current in your spinning magnet: this was one of Sturgeon's great innovations.

5. Now wind the copper wire on to the U-bolts, as shown in the diagrams.
Start with the bolt you didn't attach to the shaft. Leaving 10cm of wire free (to be connected to the battery), begin near the bend on one arm, and wind the roll of wire tightly towards the end of the arm. Aim for about 25 turns, then wind back to the starting point, giving about 50 turns in total.

6. Pass the roll of wire over to the other arm, so that the wire makes an 'S' shape between the arms, then continue winding in an identical way. Make sure the winds are in the correct direction.

7. Finish off with another good 10cm of free wire, then cut from the roll and scrape off the insulation from the ends of the wire.

8. Wind the U-bolt fixed to the broom handle in exactly the same way. Lead the loose ends along the shaft to the commutator. Scrape off the insulation from the last 1cm of each wire end. Loosen the screws slightly, and wrap one loose end around each before retightening.

9. Now mount the shaft and the fixed electromagnet. The two U-bolts should sit horizontally facing each other, with the arms almost touching.

10. Fix the mounting block firmly to one end of the baseboard (glue or

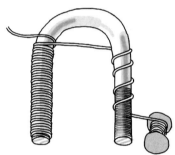

screw it). Fix the loose bolt on to the mounting block using wire staples, or a length of wire firmly screwed down. The arms of the bolt should just stick out over the edge of the mounting block. Solder the loose wire ends to copper tacks in the baseboard either side of the mounting block.

11. Fix the shaft in front of the mounting block using the L-shaped brackets. Pass screws through the bracket and into the holes drilled in the end of the shaft. At the magnet end, fix the bracket 1.5cm clear of the end of the shaft, so that the magnet can spin unobstructed by the bracket. The shaft should spin freely (Vaseline on the screws will help), without too much movement lengthways.

12. Fix the contacts (brushes) to the commutator: take about 12cm of the bronze wire, fold it in half and twist the ends together to leave a wide loop. Make another, identical. Stand

the brushes either side of the commutator, so that the loops touch the copper pipe firmly. Bend the wires at the base to adjust the height, then screw directly on to the baseboard.

13. Notice that when you spin the shaft round, the cuts in the commutator both pass a brush at the same time, so that each brush now touches the other half of the copper pipe. The brushes should never be in contact with the same half of the copper pipe at the same time.

14. Now connect the electromagnets to the batteries. Take four lengths of electrical wire, and fix crocodile clips to both ends of each. Clip one wire each to the bases of the brushes, and to the copper tacks bearing the loose wires from the fixed U-bolt. Solder if you don't have croc clips.

15. Connect the batteries, so that one battery supplies the rotating bolt

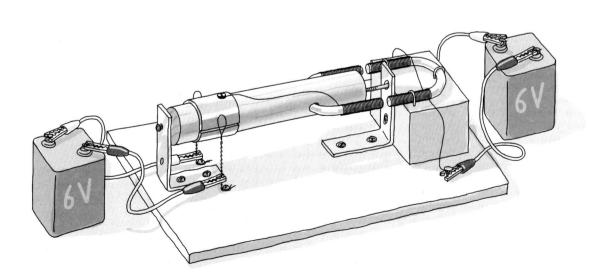

(via the brushes) and the other supplies the fixed bolt.

16. The bolts are now magnets. A quick nudge of the spinning bolt (it may favour one direction), and the motor should whirr into action!

Troubleshooting

Although this motor is incredibly simple, there are a few fiddly bits to get right. If it doesn't work at all:

- Check that the insulation is scraped from the ends of the copper winding wire to give good contact.
- Check that the brushes press firmly against the commutator.
- Loosen the mounting screws on the shaft slightly, and lubricate the screw heads.
- Try new batteries – they don't last long, or you may have shorted them.
- Check that the two halves of the commutator are not touching.

If it works, but not very convincingly:

- Make sure the arms of the bolts pass very close together, but do not touch.
- Check that the shaft is well balanced, and doesn't wobble as it spins.

WILLIAM STURGEON

Like most people, before coming across William Sturgeon, I had thought that Michael Faraday invented the electric motor. Which, of course, he did in 1821 – see Faraday's Electric Motor on page 76. However, clever though the Faraday motor was, it was never going to do any useful work. But it did show that, in principle, an electric motor could be built, and showed that getting the power to the moving bit of the motor was going to be a big challenge.

William Sturgeon shared Faraday's humble background and self-education. He was the son of a cobbler from Kirkby Lonsdale in Lancashire who joined the army to escape his cruel father. By the time he left the army he had decided to work on electricity. No one knows for sure how he became inspired, but local legend has it that one night he was taken by his father on a poaching expedition to Devil's Bridge near his home. They had to shelter under the bridge from a fierce thunderstorm, and young William was so impressed by the lightning that he formed a life-long fascination with nature's electricity.

Whatever the reason, Sturgeon had a brilliant grip of both electrical theory and practice. He made the first practical electromagnet in 1823, and then went on to invent the commutator, and thus not only the first practical electric motor but also the predecessor of all other electric motors.

Sadly, unlike Faraday, he was in the wrong place at the wrong time and did not really get any recognition for what he had done. He earned money as a popular science lecturer, but died poor and his recently discovered grave is simply marked 'William Sturgeon – The Electrician'.

WHY IT WORKS... The two bolts act as electromagnets. One end of the bolt becomes a north pole, and the other a south pole, so it is effectively a horseshoe magnet. In the motor, the two horseshoe magnets face each other. If the north pole of one is opposite the south pole of the other, then they attract each other. But in this position, the commutator switches the current in one magnet, and this reverses its polarity. The magnets now have like poles aligned, so repel each other, causing one to spin by half a turn.

At this point the commutator switches the current again, so the magnet keeps spinning – and so on.

AIR PUMP

John Barber patented this clever air pump as long ago as 1791. While it's easy to make, watch out – it can get messy!

You will need
- 2 long (e.g. 2m) plastic tubes with different diameters, so that one fits easily inside the other
- A cork, bung or lid to block one end of the larger tube
- Glue
- A valve (e.g. from a balloon pump)
- A balloon
- Water

What to do
1. Close the bottom end of the larger tube with the cork or bung, or glue a metal lid across it with strong glue such as epoxy resin.
2. Glue the valve across the top of the smaller tube. (If you have no valve you can still make and use the pump – see instructions below.)
3. Attach a balloon to the outlet nozzle.
4. To inflate the balloon, stand on a chair or table and hold the large tube upright with the closed end at the bottom. Pour in water from a hose or watering can (without the rose, of course) until the tube is a bit more than half full.
5. Carefully push the smaller tube, complete with valve and balloon at the top, down into the large tube. The balloon will begin to inflate.
6. Push the small tube down as far as you can. Then if you have a valve you can simply lift the small tube out and push it in again. Each time you push it down the balloon will inflate more.

If you have no valve
Note: The following operations need at least three hands, so you will find

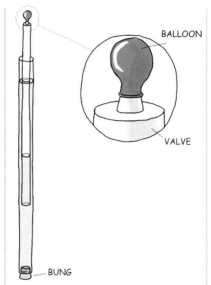

BALLOON

VALVE

BUNG

it much easier if you have an assistant! With care and cunning you can inflate the balloon as much as you want.
1. First, fit a balloon nozzle to the top of the small tube. One way to do this is to cut off the top of a pop bottle and glue it to the top of the tube. Snap a balloon over the nozzle.
2. Push the small tube down into the water in the big tube, as above. Twist and pinch the neck of the partly inflated balloon, and pull it

off the nozzle without letting any air out.
3. Lift the small tube until the bottom end is just under water. Snap the balloon on again, and push the tube down once more.
4. Repeat this process until the balloon is fully inflated.

ADAM SAYS
During the course of filming *Local Heroes* we have built about 200 demonstrations to illustrate our heroes' achievements. Some of these have been mechanical and simple, but the ones that have given us most trouble were pneumatic and hydraulic – the machines that had to be airtight or watertight.

These problems must have been much trickier in the 18th century, when plastic and rubber weren't available, and we were amazed to come across John Barber's patent of 1791.

The patent went on to describe in detail how to make a gas turbine, the kind of engine that now powers jet aircraft. The technology he had was scarcely good enough, and we're fairly sure his gas turbine could never have worked – but he did have the idea.

WHY IT WORKS... When you begin to lower the small tube into the water, the air inside the tube is trapped. As you lower the tube, the air tries to keep the water out, and so pushes it up the outer tube. This rising column of water increases the pressure of the air, which is compressed in the small tube. The water both compresses the air and traps it in. When the air pressure is high enough the balloon begins to inflate.

As you raise the small tube again, the air pressure is reduced inside, and the valve allows air in from the atmosphere, without letting the balloon deflate.

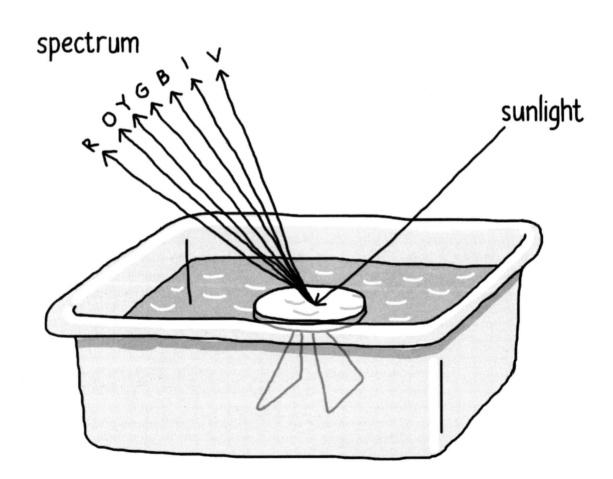

spectrum

R O Y G B I V

sunlight

CHAPTER SEVEN
LIGHT FANTASTIC

HOW TO BEND LIGHT

Light travels in straight lines, but by cunning use of reflections you can make it go round corners. One attractive way to do this is by using a light pipe.

You will need
- An old fizzy drinks bottle (preferably clear, not coloured)
- 10cm of plastic pipe (e.g. the barrel of an old ballpoint pen with the point sawn off)
- Waterproof glue
- A bowl or bucket to catch the water
- A powerful torch or a slide projector
- Some kitchen foil or black paper

What to do
1. Drill a hole in the side of the bottle, just above the curved bottom, and just big enough for the pipe to push into it. Glue the pipe into the hole with most of it sticking out, using epoxy or other waterproof glue.
2. Stand the bottle on the edge of a table with the pipe sticking over the edge. Put a bowl or bucket below the pipe to catch the water coming out.
3. Stand your torch or projector so that the beam goes through the bottle and along the pipe.
4. Make a hole in the foil or paper for the pipe to poke through, and rest it against the bottle, or wrap it half-way round, so that it cuts off most of the light that does not go through the pipe. (You don't need this foil or paper, but the effect is more dramatic if you use it.)
5. Fill the bottle with water. Either put your finger over the end of the pipe to keep the water in, or borrow a friend's finger.
6. Take the finger away and let the water run out, holding your hand in the stream of water. You should find that the light which passes down the pipe is trapped in the water stream, and bends round with it until it hits your hand, or the bowl or bucket.

More things to try
● Try tilting the bottle one way and the other, so that the pipe is not horizontal. Does the light escape, or is it still trapped in the stream?
● Try using a mirror in place of your hand. Can you hold the mirror at such an angle that the light is reflected from it and still stays in the stream?
● Using a slide projector, put in a slide with writing on it and project the light through the bottle. Use a piece of white plastic, or a plate, instead of your hand. Can you read the writing? Does it get scrambled, and if so when?

WHY IT WORKS... When light hits a water surface, it may go straight through or it may be reflected. The sun can be reflected dazzlingly by the sea or by a wet road, but sunlight can also pass through the surface, which is why you can see the bottom of the swimming pool, and sometimes the bottom of the sea. The lower the angle of the sun, the more of the light is reflected. In the light pipe the light in the stream hits the surface at a very low angle and so is trapped inside by 'total internal reflection'. Total internal reflection is the reason why light stays inside optical fibres, which now carry millions of telephone calls and billions of bits of digital information.

WHY IS THE SKY BLUE?

The answer was discovered by John Tyndall (who also invented the light pipe, opposite).
Tyndall made his own blue sky to demonstrate the effect, and this is how you can do it.

You will need
- A fish tank, made of clear plastic or glass
- A powerful torch or a slide projector
- Milk powder

What to do

1. Put the fish tank on a table and fill it with water. Set your torch or slide projector to shine through the tank from one end.

2. Sprinkle a little milk powder into the water and give it a good stir. If your tank is small you could start with as little as a quarter of a teaspoon.

3. Look into the fish tank from the side, and the water should look bluish – not very blue, since you have only a little sky. Look into the tank from the end to see the 'sun' (your torch) looking yellow.

4. Add another half teaspoonful of milk powder to the water and stir

again. The effect should be more obvious, and the sun will look reddish, as at sunset.

DID YOU KNOW?

Liquid oxygen is blue, and oxygen gas may also be very slightly blue, although in small amounts it looks colourless. We have miles of it over our heads, and so it is possible that oxygen gas also contributes to the colour of the sky.

The brilliant Irish scientist and mountaineer John Tyndall succeeded Michael Faraday at the Royal Institution in 1867.

WHY IT WORKS... Tyndall noticed that when he shone a beam of light through filtered air he could not see the beam. He realized that the only reason he could usually see beams of sunlight shining in through a window and across the room was that the air was full of tiny particles of dust.

When you first shine your light through the water in the fish tank you may not be able to see the beam from the side, if the water is really clear. However, when the water is full of milk particles you can see the beam from the side, because the particles scatter the light. When a ray of light hits a particle it can bounce off in any direction; so the light is scattered in all directions.

Small particles scatter blue light more than red light. Therefore when you look from the side of the tank you see scattered light that is slightly blue, and when you look from the end you see the lamp yellow or red, because blue light has been scattered away, and what is left is the rest of the spectrum, which is yellow or red.

If the air above our heads were absolutely clear the sky would be black, but because it is full of particles of dust and water the blue of the sun's light is scattered down to us, and the sky looks blue.

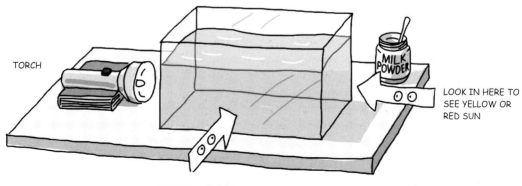

TORCH

MILK POWDER

LOOK IN HERE TO SEE YELLOW OR RED SUN

LOOK IN HERE TO SEE BLUE SKY

NEWTON'S RAINBOW

Isaac Newton was one of the greatest scientists of all time. In a single year he invented calculus, cracked some major mathematical problems, worked out for himself the basis of gravitational attraction and solved the mystery of the colours of the rainbow.

You will need

- A washing-up bowl or a large saucepan
- A small mirror – ideally a shaving mirror on a wire stand
- You'll also need a sunny day. You can make a spectrum using a bright lamp such as a security light or a slide projector, but sunlight is much the best.

What to do

1. Find a room where the sun shines in through a window. Draw the curtains, leaving only a small hole for the sun to shine through. Newton said, 'I made a small hole in my window shuts,' but you don't actually have to cut a hole!

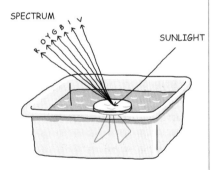

SPECTRUM

ROYGBIV

SUNLIGHT

2. Place the bowl or saucepan in the patch of sunlight, on the floor or on a table, and half fill it with water. Put the mirror in the bowl so that the sunlight goes through a few centimetres of water before hitting the glass and is then reflected upwards.
3. You should then, by adjusting the position and angle of the mirror, be able to get a spectacular spectrum of colours on the ceiling, if it is white, or on a piece of white card if the ceiling is coloured.
4. Another way to view the spectrum is to hold a piece of tracing paper above the bowl and look down from above it.
5. If you get a white patch with coloured edges then your hole is too big. Make a smaller hole and let in a narrower beam of light.
6. The spectrum will be most brilliant if the room is dark; so do your best to keep out all light apart from the narrow beam of sunlight that hits the mirror.

More things to try

● How many colours can you see in your spectrum? Newton claimed he could see seven distinct colours – red, orange, yellow, green, blue, indigo and violet. However, the colours all run into one another, and most people are not convinced there are two colours beyond blue. What do you think?
● Try to get a good photograph of your spectrum. To show the full brilliance of the colours you must make sure you turn off the flash gun and eliminate all stray light in the room.
● You could try what Newton described as the 'experimentum crucis' (the crucial experiment), but beware: this needs a second prism, and is very tricky – indeed so tricky that Robert Hooke said it didn't work, and started a furious dispute with Newton.

THE 'CRUCIAL EXPERIMENT'

You will need

- A piece of cardboard
- A second prism (if you can't get hold of one, you can make your own; see opposite)
- A protractor

What to do

1. Cut a slit about 2mm wide in the piece of cardboard. Prop or stand the cardboard at least 50cm above the mirror so that it intercepts the spectrum. Adjust the position of the cardboard until the slit lets through light of only one colour – say red.
2. Clamp your prism above the slit so that the red light passes through it and is refracted.
3. Now the 'simple' part of the 'experimentum crucis' is to measure the angle through which the light is refracted. Hold the protractor against the end of the prism, measure the angle through which the beam of red light is bent and make a note of it. (See diagram above right.)
4. Move the cardboard to let through orange light. Measure the angle of refraction and write it down.
5. Continue the process for each colour. You will discover which colour is refracted most and which least. This is the basis of Isaac Newton's discovery that white light is split by the prism into its separate colours.
   The second part of his crucial experiment is more difficult to make truly objective.
6. Replace the slit so that it lets only red light through. Look at the ceiling

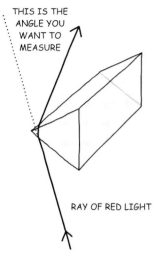

THIS IS THE ANGLE YOU WANT TO MEASURE

RAY OF RED LIGHT

(or your tracing paper) and see whether the red light is split by the second prism into different colours. Is the patch of refracted light pure red, as Newton said, or are there other colours too, as Hooke said?

Before Newton's time, people thought prisms made white light stained. Newton's experiments showed that in fact white light from the sun is a mixture of all the colours of the rainbow, and that the prism merely separates them. Therefore if you can isolate a pure colour from the spectrum, a prism should not be able to separate it into any more colours. However, to make an absolutely pure colour you would have to use an infinitely narrow slit, and the narrower the slit the fainter the light that comes through, which makes it harder and harder to tell what colour or colours there are.

In other words the observation becomes subjective: you can believe you see what you want to see – which is why Hooke and Newton disagreed.

HOW TO MAKE A PRISM

RATING: ✳ EASY!

You will need
- Some clear plastic (e.g. Perspex or acrylic)
- Waterproof glue
- Cold, boiled water

What to do
1. Cut the plastic into three pieces about 2 x 6cm. Glue the long edges together to make a triangular tube (like a Toblerone box).
2. Cut two more pieces of plastic for the ends, and glue one of them on. When the glue has set, stand the tube on its closed end and fill it with cold water that has been boiled (to remove dissolved air so that fewer air bubbles will form later).
3. Carefully glue on the top, ideally leaving the tube full of water.
4. You might find it easier to leave a small gap at one corner, let all the glue set, and then carefully top up with water before sealing your prism with a blob of glue.

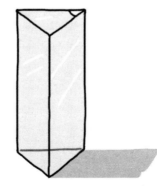

DID YOU KNOW?
To remember the colours of the spectrum, there are two simple mnemonics: the word ROYGBIV, standing for Red, Orange, Yellow, and so on, and the sentence Richard Of York Gained Battles In Vain. If you can see only six colours, you might like to make up your own mnemonic to recall them.

NEWTON'S PRISMS
The first scientific paper Newton published was a letter to the Royal Society, in which he wrote, 'In the beginning of the year 1666, I procured me a triangular glass prism, to try therewith the celebrated phenomena of colours.' He went on to describe his refraction experiments, and in passing how to make a reflecting telescope! He probably bought his prism at Stourbridge Fair near Cambridge, but his letter is a bit disingenuous, since for his 'experimentum crucis' he suddenly introduces a second prism, and it's hard to believe he went all the way back from Woolsthorpe to Cambridge (about 75 miles) just to buy another prism; presumably he bought two to start with!

This letter reads as though Newton had just finished doing his experiments, but in fact he wrote it in 1672, five or six years later, by which time he had become Lucasian Professor of Mathematics at Cambridge. So he had had several years to think about the results before he committed himself to paper – and even then Hooke said he had got it all wrong!

FIRE FROM ICE

While trapped in the ice in Greenland, Whitby whaler William Scoresby Junior did scientific experiments. One of the most intriguing was lighting sailors' pipes with a lens made of ice.

DANGER FROM ICE BURNS FROM TOUCHING ICE / CARE WITH BURNING PAPER

You will need
- A round ceramic or glass bowl you can put in the freezer
- Clingfilm
- Tap-hot water
- Cold water
- Gloves
- Paper to burn
- Metal tray or similar to put the paper on

While the bowl must be round, the diameter doesn't matter. Start with one about 15cm across. You'll also need the use of a freezer and a sunny day.

What to do
1. First make a mould for your lens. Half-fill your bowl with hot water, then immediately cover the bowl with

COVER WITH CLINGFILM

WATER COOLS, AIR SHRINKS

clingfilm. As the water cools, it cools the air which will contract, pulling the clingfilm into the bowl in a nice smooth curve. (You may have to experiment with different temperatures of water.) This is your mould.

2. When the mould is cool, fill with cold water, and place in your freezer until it has frozen solid. Note that the water in the bowl will also freeze, and because water expands when it turns to ice, you should be sure that you're using a bowl that is strong enough.

In practice something like a heat-resistant glass bowl used in the kitchen will be fine.

3. When it has frozen, remove the mould from the freezer, and take off the clingfilm. You may have to run water over the surface to do this. Use gloves to handle the ice lens.

4. Take the lens outside, and see if you can use it as a 'burning glass' by focusing an image of the sun on to a bit of paper. Make sure the paper is somewhere safe in case it does burst into flames.

You may find that the lens is not very transparent. Try putting it into a bowl of cold water from the tap. As the ice warms up, it should clear. The surface will also become smoother as it melts a little. Scoresby actually moulded his lenses by hand – you could try this.

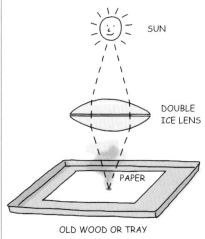

SUN

DOUBLE ICE LENS

PAPER

OLD WOOD OR TRAY

More things to try
● Make two lenses and stick their flat sides together, to make a double convex lens like a magnifying glass.
● Try making flatter or more curved lenses – what difference does this make? Does it affect the distance you have to hold the lens from the paper to get a clear image of the sun?

ADAM SAYS
Really clear ice is made by freezing water very slowly and stirring it as it freezes. Ice-sculptors freeze their ice over three days at temperatures only just below freezing.

KALEIDOSCOPE

You've probably had a kaleidoscope in your Christmas stocking at some time. But how was it invented? By accident, of course, by optical scientist Sir David Brewster.

THE TWO-MIRROR KALEIDOSCOPE

RATING: ✳ EASY!

DANGER FROM GLASS, IF USING

In 1816, Brewster was working in the lab, making an optical trough. This was a water-filled triangular trough with the sides made of mirrors. He accidentally dropped a huge blob of glue on to the place where the mirrors joined – and noticed something strange. The shape of the blob was multiplied up in a peculiar but geometric way. This was the first kaleidoscope.

It might have been an accident, but he had the wit to recognize that he had stumbled on something special that would appeal to the public. If he had bothered to patent it he might have become rich as well as famous. He didn't – and the idea was stolen by others. Here you can make both Brewster's original two-mirror kaleidoscope, and the better-known three-mirror variety.

You will need
- Mirrors
- Gaffer tape

It's much safer to use plastic mirrors if you can get them. Try craft shops – or some DIY and furniture stores sell them as decorative mirrors.

Otherwise, mirror tiles are good (try in DIY stores again). If using glass mirrors, either buy complete mirrors or have them cut professionally with polished (blunted) edges. Large mirrors are more fun – say 30cm square.

What to do
1. Place two mirrors together, reflective surfaces facing in; fix them together with gaffer tape as a hinge. Now stand your kaleidoscope on something interesting – this book, for instance – and see what happens as you change the angle of the mirrors. View the object by getting your eye as close as you can to the place where the mirrors join.
2. The image in the kaleidoscope is always interesting, but particularly pleasing when the object is repeated a whole number of times, as happens when the angle between the mirrors exactly divides into 360. So you should get neat patterns with mirrors at 120, 90, 72, 60 and 45 degrees – which go into 360 3, 4, 5, 6 and 8 times. But do you also get a pattern at 51.4 degrees – 360 divided by 7?

THE THREE-MIRROR KALEIDOSCOPE

Take your two-mirror kaleidoscope and add a third mirror, making a tube with angles of 60 degrees. Although the three-mirror version gives an all round sensation, I think it's more boring than its simpler predecessor, where the fun comes from trying out different angles. Perhaps that's why all modern three-mirror kaleidoscopes seem to have bits of coloured plastic to shake in front of them, to give you something interesting to look at.

DAVID BREWSTER

He tried to be a vicar, but was too nervous to deliver sermons. Instead, he became a celebrated contributor to magazines and encyclopedias about science, before becoming a professional scientist.

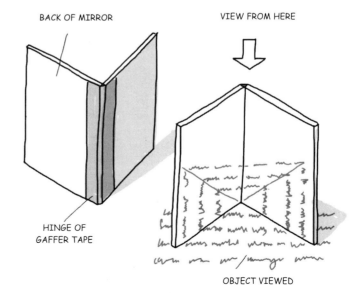

BACK OF MIRROR

VIEW FROM HERE

HINGE OF GAFFER TAPE

OBJECT VIEWED

PINHOLE CAMERA

You might choose to make a simple pinhole camera to view the world (upside-down), or you could make a more elaborate one to take real photographs.

SIMPLE VIEW CAMERA

RATING: ✳ EASY!

You will need
- A shoe box or any shallow cardboard box, or a biscuit tin
- Knife or scissors, or hammer and nail
- Tape
- Kitchen foil
- Pin or needle
- Tracing paper, greaseproof paper or baking parchment

What to do

1. In the middle of the bottom of the box or tin, make a small hole, say 3–5mm across, using a knife or scissors for the cardboard, or a hammer and nail for the biscuit tin.

2. Over this hole, tape a piece of kitchen foil, and use a pin or needle to make a small hole in the centre of this. The point of using the foil is that you can make a nice clean round pinhole in it; also you can easily change the size of the pinhole without needing a new box.

3. Over the open top of the box or tin tape a piece of tracing paper.

4. Your pinhole viewer is now ready for use. Try standing it on a windowsill with the pinhole pointing outwards on a bright sunny day. You should be able to see an image on the tracing paper of the world outside. You will notice that it is upside-down and reversed from left to right. This proves that light travels in straight lines.

5. You will be able to see this image much more easily if you make the room dark — draw the curtains and switch off any lights. Then the image will look much brighter. The Victorians

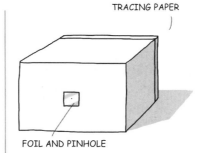

TRACING PAPER

FOIL AND PINHOLE

called such a darkened room a camera obscura.

The pinhole viewer should also work well at night, when you will be able to see images of street lamps, or lit windows in other buildings. Again, make sure no light falls on the tissue paper except through the pinhole.

More things to try

● Instead of leaving it on the windowsill, take your camera outside during the day, and try it in the street or out in the country. Keeping light off the back of the screen is important; you should take along a piece of black cloth – perhaps a coat – to put over your head to keep light off the screen.

● Try using pinholes of 0.5mm, 1mm and 2mm, and see which gives the best results.

How big should the pinhole be?

Imagine using the viewer at night, and looking at a distant street lamp. It should appear as a point on your screen. If the pinhole is 1mm across, it will appear as a 1mm circle. If the pinhole is 2mm across, the lamp will appear as a 2mm circle. So the bigger the pinhole, the fuzzier the image will be; or the smaller the

pinhole the sharper the image. However, changing the pinhole from 1mm to 2mm lets in four times as much light; so your image will be much brighter with the bigger pinhole.

PINHOLE CAMERA TO TAKE PHOTOGRAPHS

RATING: ✳✳✳✳ DIFFICULT

It's the materials that can be difficult to deal with, rather than the actual construction of the camera. There are various ways to tackle this; we suggest using instant-print film, which is not only good fun but provides rapid feedback about exposure, and so gives a good chance of taking useful photographs. The drawback is that both the back and the film are expensive. Alternatively, if you have a darkroom you could try using photographic paper. You will have to load each sheet in the dark, and when you process it you'll have a negative image.

You will need
- Glue or sticky tape
- Black paper
- A cardboard box, ideally 12 x 17cm and, say, 15–25cm long (or make your own box to fit the Polaroid back)
- A Polaroid back suitable for a medium-format camera
- Black tape
- Kitchen foil
- A pin or needle
- A pack of instant-print film to fit in the Polaroid back

What to do

1. Glue or tape black paper to all

cardboard surfaces that are going to be inside the box. This will cut down stray reflections.

2. Either fit the Polaroid back into one end of the cardboard box and tape it in place so that the hinged side opens outwards and the metal slide is on the inside, or make up the box with the Polaroid back as one end by taping pieces of cardboard together.

3. Stick black tape along all the outside edges of the box, to keep out stray light.

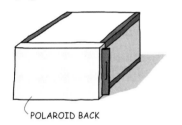

POLAROID BACK

4. In the centre of the front end, cut a small hole about 3mm across and tape a piece of kitchen foil over it. Using a pin or needle, make a pinhole in the foil about 1mm across.

5. Your camera is now ready for use. You don't have to focus it; as long as the pinhole is small enough everything will be in focus. However, you do have to guess the exposure time. If you have a camera with an exposure meter, you can get some idea by pointing the camera at the scene you want to photograph and seeing what exposure time it says, allowing for a film speed of ISO 100 and an aperture of f/32.

6. With fairly bright sunshine you might like to try an exposure of 60 seconds. If the print you get is overexposed – too bright – try 15

seconds, and then if it is still too bright try 4 seconds. On the other hand if the first one is too dark, try 4 minutes, and so on.

7. Insert the instant-film pack into the Polaroid back, shut the back, and pull out the long black paper protecting strip.

8. Take the camera to your chosen location and rest it on a steady support, such as a table or a wall. Hold a piece of black card over the pinhole.

9. Remove the metal slide protecting the film pack. Take the card away from the pinhole for the length of your chosen exposure.

10. Cover the pinhole with the black card, and replace the metal slide.

11. Pull out the exposed film and allow it to develop, according to the instructions on the pack.

12. If the resulting picture is too dark or too light, take another picture with a corrected exposure time.

More things to try

● Because the pinhole camera is small and doesn't let in much light, a pinhole camera always needs long exposures. This means you cannot use it for action pictures, but you can put

these long exposures to other uses. Take a photograph of a busy street or market place, and you will find that most of the people have disappeared because they move too quickly to show up on the film. A picture of a waterfall or rapids will show the water as a long blur like a sheet of silk. To take a portrait of a person you will have to get your subject to sit very still, but if you are lucky – and patient – you may get an attractive portrait.

DID YOU KNOW?

The pinhole camera was invented about a thousand years ago in Egypt by an Arab scientist called Abu Ali Mohamed Ibn al-Hasan Ibn al-Haytham, or Alhazen for short. However, he had no instant-print film; so you can do better than he did!

One of his less successful experiments was his offer to Caliph al-Hakim of Egypt to solve the problems of the flooding of the Nile by building a dam – only to discover the Nile was far too wide to build a dam with the equipment he had! To avoid punishment by the tyrant Caliph he pretended to go mad until the Caliph died in 1021.

WHY IT WORKS... Light always travels in straight lines. If it didn't, we would be able to see round corners and sunlight would not make shadows, because everything would be lit by the sun all the time.

Alhazen (see above) made a camera obscura – a dark room with just a pinhole in the shutter over the window – and showed that at night the five lamps outside made five spots of light on the far wall of the room.

By holding up his hand in the middle of the room he could intercept any one light beam and so remove whichever spot he wanted, and he demonstrated that he could intercept that beam at any point on the straight line between the pinhole and the spot on the wall. Therefore the light from each lamp must be passing in a straight line through the pinhole to the far wall.

CAMERA LUCIDA

This device is one of our demonstrations that has proved most popular with viewers. Although it was shown in an early *Local Heroes* series we still get letters asking for the thing that helps people who can't draw to draw. So here it is.

DANGER FROM GLASS

The camera lucida was invented by William Hyde Wollaston of Dereham in Norfolk. Camera lucidas were once on sale, and similar devices were made twenty years ago for kids, but I don't think you can buy them now. So if you want one, you'll have to make it.

There were in fact various sorts of camera lucida. They all involved one main idea: that you can combine an image of the thing you want to draw with an image of the paper you are drawing it on. The simplest camera lucida is just a sheet of glass set at 45 degrees. You could make one by gluing the glass into a slot in the cork of a wine bottle. It works perfectly well, by allowing you to see the paper *through* the glass together with the thing you are drawing reflected *off* the glass. However, the subject will appear upside-down, and I suggest you try the more sophisticated version below.

You will need

- Handbag-type mirror – about 50mm x 75mm, or smaller (a much smaller one also works very well)
- Similar size piece of glass or shiny plastic

IF YOU ARE USING GLASS, IT MUST HAVE POLISHED (SMOOTHED) EDGES. FRESHLY CUT GLASS IS ONE OF THE SHARPEST THINGS YOU CAN FIND AND THIS IS GOING TO BE CLOSE TO YOUR EYE.

- Paper and pencil
- Protractor (optional – you can get away without it)
- Plasticine
- Sticky tape
- Epoxy glue
- Piece of dowel or knitting needle
- Wine bottle or similar – optionally with sand or water to weigh it down
- Cork for above

What to do

1. The glass and mirror have to be fixed together along the long edge at an angle of 135 degrees.

2. The angle of 135 degrees is easier to think of as 45 degrees from a straight line – 180 minus 45 is 135. So the first step is to draw two lines at 135 degrees using either the protractor or by first drawing a straight line, then another at 45 degrees to it. You can of course make your own 45-degree template by folding the square corner of a piece of paper in half.

3. Stand the mirror and glass on the template, so that they meet and are at 135 degrees. So that they don't move, use Plasticine to hold them in place on the piece of paper and perhaps a bit of sticky tape over the edges at the top. Glue by dribbling mixed epoxy down the back of the join. The glue should be mixed according to the instructions. Epoxy is usually a two-pack adhesive, and you use equal parts of 'adhesive' and 'hardener'. Mix thoroughly, and use within the stated time – usually a quarter of an hour unless you've bought the five-minute epoxy.

4. I have sometimes had problems keeping all the parts together, and once they are fixed in position you might consider a second application of epoxy. Avoid getting too much adhesive on to the front viewing surface.

5. When the glue is set, fit the shaft – a piece of dowel or knitting needle to the glass and mirror. You now have to decide which eye you are likely to

GLASS

MIRROR

use. The camera lucida is used with the glass up, and closest to the eye. So if you use your right eye, you'll want the shaft out to the right, the reverse for the left eye. However, you also need to think about which hand you draw with, because it is easier to have the camera lucida to your left if you are drawing with your right hand. If you find that your hand and eye don't match, my advice is to go with the hand – it is much harder to draw with the wrong hand than to look with the wrong eye.

6. Holding it in place with tape, fix the mirror and glass assembly to one end of the shaft with more epoxy glue.

Mounting the camera lucida

1. If you have some sort of clamp stand, use that. Or you may be able to fix the camera lucida to one of those sprung desk-lamps for a really posh drawing device. However, I tend to have a lot of empty wine bottles hanging about and one of those is fine.

2. Put some sand or water in the bottom to weigh it down if you like, then push the knitting needle through the cork and push that into the bottle. And that is it. Younger artists may find a wine bottle uncomfortably high, in which case you'll have to use a smaller bottle.

3. Set up the camera lucida next to a large sheet of white drawing paper, in front of a well-lit scene. Make sure you are sitting comfortably.

Using the camera lucida

1. Place your eye close to the glass. Look through the glass directly at the drawing paper: you will also see the image of your subject, which has been reflected off the mirror and the surface of the glass.

2. Trace round the image you see, which will be the right way up having been reflected twice. It really is as easy as that, though the technique takes some getting used to. In particular, you can't simultaneously focus on, say, a tree at 100 metres and a piece of paper at 20 centimetres. Nevertheless, for landscapes and buildings I have found the simple camera lucida to be surprisingly effective. It really is a device that helps people who can't draw to draw.

WOLLASTON'S CAMERA LUCIDA

Wollaston addressed a problem you may have with this design. Although the image is now upright, you can't control the brightness. This is a problem if the subject image is so bright or dark that it obscures the drawing. I found that I could draw a sunlit house quite well, but got into trouble with the darker bushes and trees.

Wollaston had two modifications to sort this out. Instead of looking through a plain glass, Wollaston used a partially silvered mirror, with the silvering in the shape of a triangle. You could choose to look through the thinner or thicker end of the triangle of mirror, and thus you saw more or less reflected light from the subject. Because it was so close to your eye, you weren't aware of the shape of the mirror.

A different approach was his 'split pupil' design, where you used two mirrors at 135 degrees, but placed your pupil right at the upper edge of the mirror. Part of your pupil looked at the mirror, part directly at the paper, so mixing the image. With both designs (half mirror, split pupil) users became expert at moving the eye as they were drawing to adjust the relative brightness of subject and drawing. I found the split pupil design impossibly hard to use, but the triangular half-mirror was really good.

Another problem is that each time you look up at your subject or take a break, image and drawing can go out of alignment. Perhaps a nose- or chin-rest would solve the problem?

CHAPTER EIGHT
SCIENTIFIC INVESTIGATION

GALILEO AND GRAVITY

Galileo Galilei was a brilliant Italian scientist of the Renaissance who, among other great achievements, worked out the way in which things fall. Galileo managed to describe falling in the language of mathematics.

THOUGHT EXPERIMENT

RATING: ✳ EASY!

Ever since the time of Aristotle, people had believed that heavy things fall more quickly than light things. Indeed, if you think about it you can convince yourself that this must be true; for example, that a whole brick will fall faster than a half-brick.

Galileo is supposed to have dropped large and small objects from the top of the Leaning Tower of Pisa in 1590, and showed that they hit the ground at the same time. But you don't have to climb the Leaning Tower – you can do the experiment in your head. Einstein called such an experiment a 'gedanken' experiment, or thought experiment.

You will need
- Just your head!

What to think
1. Imagine dropping a whole brick and a half-brick at the same moment from a high tower. You may think the whole brick will hit the ground first. Now imagine repeating the experiment, but noticing just before you let go that the brick has a crack running through it. This cannot make any difference to the result.
2. Now imagine doing the experiment for a third time. The brick has come in half at the crack, but you tape the two halves tightly together so that it behaves just like a whole brick.
3. Do the experiment for a fourth time, but tie the two pieces loosely together. As the two halves separate, will they suddenly fall at a slower

rate? No; they must fall at the same speed however far apart they are – and therefore a half-brick falls at the same speed as a whole brick. Note that this applies only to objects that are similar in shape and density – i.e. we're not comparing boulders and feathers here!

ROLLING EXPERIMENTS

RATING: ✳✳ MODERATE

One of Galileo's most important pieces of work was to show how falling objects accelerate, and this is roughly how he did it.

You will need
- Approx 2m length of plastic gutter, preferably semicircular in section (if you have two pieces of gutter, so much the better)
- A table, ideally nearly as long as the gutter
- Wood, bricks or books to jack up the table
- Marbles, preferably at least one small and one large
- An accurate timer of a few seconds – e.g. a metronome, or a ticking watch or clock (or you could make your own – see page 126)

What to do
1. Lay the gutter on the table. Tilt the table up about 10cm at one end, by putting blocks of wood, bricks or books under two legs.
2. If you have two pieces of gutter, you can easily test whether big things fall faster than little ones. Put a big marble at the top end of one slide (piece of gutter) and a little one at

the top of the other. Release them at the same time, and see whether they reach the other end together. You are effectively dropping them, but using the shallow slope to slow their fall.
3. However, measuring acceleration is more tricky. Place a marble at the very top of the slide. Listen to the watch or timer. As you hear a tick, let the marble go. As you hear the next tick, make a mark on the slide opposite where the marble is. You could use a pencil, or a piece of tape, or a paper-clip.
4. Repeat this at least three times, making another mark each time. Judging exactly where the marble is when you hear the tick is difficult, but you should be able to get a fairly precise average position after four or five runs.
5. Now start the marble at the top again, but note where it is at the second tick. Repeat this at least three times, making a mark each time.
6. Repeat the process for the third tick – and for as many ticks as there are before the marble goes off the end of the slide.
7. Measure the distances from the top of the slide to the position at the first tick, from the first tick to the second tick, from the second to the third, and so on.
8. Write these distances in a table like the one shown. The actual distances will depend on the slope of your table and the frequency of your ticks, but you should find that they increase – each distance is bigger than the one before. This demonstrates that the marble is accelerating as it rolls down the slope.

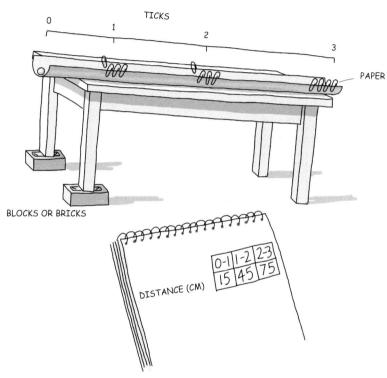

TICKS

0
1
2
3

PAPER

BLOCKS OR BRICKS

DISTANCE (CM)

0-1	1-2	2-3
15	45	75

Galileo showed that the ball has constant acceleration as it rolls down the slope. This is the acceleration due to gravity. No one had ever treated mechanical movement in this mathematical way before.

9. To find out whether your marble has constant acceleration, you need to work out the differences between the distances covered in each interval between the ticks. In the first interval our marble went 15cm; in the second it went 45cm, which is 30cm more than in the first. In the third interval it rolled 75cm, which is 30cm more than in the second. So from that we can tell that our marble had a constant acceleration of 30.

If our ticks were one second apart, this would be a constant acceleration of 30cm per second.

However, this is an idealized experiment, and you may find that because of friction your acceleration is not quite constant.

More things to try

● Try the same experiments with a marble of a different size. Do you get the same result?

● Try varying the angle of the slope by changing the size of the blocks. What happens then?

● For any set of results, try plotting a graph of the total distance travelled against the square of the number of ticks. So for our results above, plot the points 0,0; 15,1; 60,4; and 135,9. What is the shape of the graph, and what does this show you?

DID YOU KNOW?

Galileo overthrew the ideas of Aristotle and the ancient Greeks, who had believed that stones fall to earth because that is where they belong naturally, and that as they fall they fall more quickly because as they get closer to home they move more jubilantly! Galileo said that the book of Nature is written in the language of mathematics, and his rolling experiments showed that natural events could be described mathematically. He died in 1642, the year in which Isaac Newton was born, and it was Newton who worked out the laws of motion.

WHY IT WORKS... The acceleration is constant because acceleration is caused by force, which in this case is the force of gravity – the weight of the marble pulling it down the slope. Because the force of gravity is constant, the acceleration is constant. In free fall without air resistance, objects near the Earth's surface fall with an acceleration of nearly 10m per second, and fall 5m in the first second, 15m in the second and 25 in the third.

THE MOONS OF JUPITER

The first person ever to see the moons of Jupiter was the Italian astronomer and mathematician, Galileo. He heard about the Dutch invention of the telescope in May 1609 and quickly built his own instrument, which you can do too.

HOW TO SEE THE GALILEAN MOONS

RATING: ✳✳✳ CHALLENGING

You will need
● Either a fairly good pair of binoculars or a small telescope (how to make your own, below), and a firm support for them

A tripod is an ideal support, but you may be able to get away with placing your binoculars on a wire support or a soft sweater on a table. You can't hold them steady enough with your hands. You also need to be supported yourself in a comfortable chair (e.g. a reclining garden seat), and you have to choose a clear night when Jupiter is visible.

What to do
1. You can find out when Jupiter is due to be visible by looking in the quality newspapers on or near the last day of the month, when they tell you what is visible for the next month. (Or phone Science Line on 0808 800 4000.)
2. It's not too difficult to locate Jupiter – it's bright, usually the brightest thing in the night sky after our moon and Venus. Venus is always close to the sun, and therefore visible only soon after sunset or soon before sunrise. If, as the sky gets dark, the first heavenly object you see is not near the horizon, then it is probably Jupiter.
3. Wait for a clear night, and either go outside or find an open window from which you can see Jupiter.
4. Prop up your binoculars or telescope and look at the planet.

THE MOONS OF JUPITER

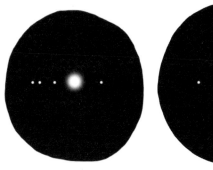 

ONE NIGHT   A FEW NIGHTS LATER

Finding it takes a bit of practice, but remember it is bright, and it looks big – more than a pinpoint of light.
5. Look for three or four pinpoints of light close to Jupiter and in a straight line with the planet. These are Jupiter's moons. There are more, but only four of them are big enough to see with binoculars or a small telescope. This is what Galileo saw, and it was a turning point in modern science.
6. Draw a diagram of the positions of the moons and Jupiter.
7. Wait for a day or two, and then repeat the process. Draw another diagram, and you will probably find the moons are in different positions.
8. Sometimes you can see all four moons; at other times only two or three. They revolve around Jupiter, but because we are looking across their orbits they seem to move to and fro in a straight line.

MAKE A TELESCOPE
The level of difficulty for this project depends on just how you do it: it's easy to make a poor telescope, but very difficult to make a good one. In about 10 minutes I made a primitive telescope through which I could read a car number plate at 50 metres, but I could not see Jupiter's moons.

You will need
● A large lens (magnifying glass) with a long focal length
● A small lens with a short focal length
● Cardboard or plastic tubes (optional)
● Sticky tape

You can buy cheap magnifying glasses at stationery shops, or good lenses from an optician.
To make a really good telescope you need achromatic lenses, otherwise you will see coloured fringes.

What to do

Before you do anything, remember:

NEVER LOOK AT THE SUN THROUGH A LENS – YOU CAN BLIND YOURSELF IN SECONDS.

It's harmful enough looking at the sun with the naked eye, but doing so through binoculars or a telescope will destroy your sight.

1. Bearing this warning in mind, find the focal length of each lens by looking through it at the skyline (NOT THE SUN) and moving it slowly away from your eye until the image suddenly goes upside-down, so that the sky is at the bottom. When the image flips, the lens is just one focal length away from your eye. Often lenses of short focal length are small, while lenses of long focal length are large.

2. The simplest way to use your telescope is to hold the small lens close to your eye and hold the long lens at arm's length, lining them up on the thing you want to look at. Try looking at a tree or a house on a sunny day, or at the moon by night.

3. Move the big lens closer and further away until the image comes into focus. It will be upside-down. You should be able to get a magnification of two or three times.

4. If you are enthusiastic, you can mount the lenses in cardboard tubes, as shown in the diagram. This will allow you to point the telescope more easily, and to focus accurately and then keep the focus. However, you don't need tubes for the telescope to work.

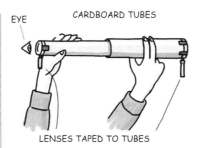

EYE CARDBOARD TUBES

LENSES TAPED TO TUBES

5. Try to find two long cardboard tubes of slightly different diameters, so that one slides inside the other. The total length must be greater than the sum of the focal lengths of your lenses, but with average magnifying glasses two 30cm tubes should be fine. Tape the big lens to the end of the fatter tube and the little one to the thinner tube.

6. If you can make a telescope good enough to see the moons of Jupiter you have done a brilliant job!

GALILEO AND HIS MOONS

When Galileo looked through his telescope at the night sky he was amazed. He quickly realized that the Milky Way and nebulae were just vast clusters of stars. He looked at the Moon and saw mountains and craters, and this was scary because the establishment view was that all heavenly bodies were perfect spheres, and never changing. Clearly the Moon had changed during the course of its history – just like the Earth – and was not a perfect sphere.

Then on 7 January 1610 he looked at the planet Jupiter and saw four moons, now called Io, Callisto, Europa and Ganymede. Subsequently, more moons have been discovered around Jupiter (the planet has at least 16), but the first four identified by Galileo are still known as the Galilean moons. As he watched over the next few nights, the moons changed positions around the planet – and the world of science was changed for ever.

Galileo was already fairly sure that the planets Venus, Earth, Mars, Jupiter and so on revolved around the sun, but when he saw the movements of the moons of Jupiter he realized that no one could still logically believe that the Earth was the centre of the universe. Clearly the moons were going round Jupiter, and therefore they were not going round the Earth.

But the men of the church insisted that the Bible must be right, and that the Earth must be the centre of everything. When he first wrote about Jupiter's moons, Galileo was sternly rebuked by the church, and told not only that he must not write about such things, but that he must not even believe them. This was the first serious rift between science and the church.

For his beliefs, Galileo was put on trial before the Inquisition in Rome and forced publicly to deny his beliefs about a sun-centred universe. He had to live the rest of his life under house arrest, so dangerous did the church view his beliefs. If only Galileo could know that spacecraft such as Voyager, and one named after Galileo himself, have now flown past Jupiter and its moons and sent pictures back to Earth.

JELLY TOWERS

This is both a challenge and a genuine experiment: you could break the world record for the height of a jelly tower. Be warned – you need patience and a huge amount of jelly and pasta. And clearing up the mess at the end is hard work!

RATING: ✱ **EASY!**

You will need
- A flowerpot or bucket at least 30cm high, with holes in the bottom
- Sticky tape or clingfilm
- Oil or Vaseline
- A container to measure volume
- Jelly (any flavour or flavours you like) – at least 20 packets, but measure your pot first (see below)
- Approx 3kg pasta (perhaps long curly ones, such as fusili)
- A tray or piece of board big enough to cover the top of the pot
- A camera to record the result

Why the pasta, do you ask? When we tried making a giant jelly, 30cm high, and turned it out of its container, it collapsed in a sticky heap. We concluded that before it can be used as a structural material, jelly needs to be reinforced. Just as concrete is cement reinforced with aggregate (sand and gravel), so we discovered that jelly can be reinforced with pasta, and we succeeded in making a pasta-reinforced jelly tower about 25cm high. The question is: Can you do any better?

What to do
1. Tape over the holes in the bottom of your flowerpot or bucket from both the inside and the outside, or cover the bottom with clingfilm, or use both tape and clingfilm.
2. Stand it in the sink or outside, where leaks don't matter, and measure its volume by filling it with water using a container of known

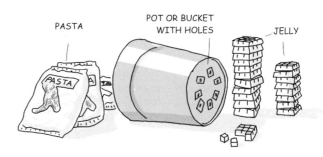

PASTA   POT OR BUCKET WITH HOLES   JELLY

volume. For example, if it takes 30 half-litre milk-bottles full of water to fill the pot, then it holds 15 litres. This tells you how much jelly you will need; a packet of jelly usually makes 600ml, but you might want to make it a bit stronger than normal.
3. Empty out the water, but leave the tape or clingfilm in place. You might also like to grease the inside slightly with oil or Vaseline, to help stop the jelly sticking to it.
4. Mix up your jelly according to the instructions on the packet (or you

POUR JELLY MIXTURE OVER PASTA

might like to make it extra strong – ours was about one-third stronger than normal). This should strengthen the structure, but it means you will need more jelly. You can make it in batches if necessary.
5. Tip the dry pasta into the pot, but make sure it doesn't come within about 5cm of the top, since it expands a little when it gets wet.
6. Pour the warm jelly mixture over the pasta in the pot. Continue adding warm jelly until the pot is full to the brim.
7. Leave the jelly to set in a cool place. This takes a long time because of the large volume of jelly. You will need to leave it for many hours, and preferably overnight. Do not try the next stage until the surface is set good and hard.

Get your camera ready
1. Place the tray or board on top of the pot. Quickly and carefully turn the whole thing over, and stand the pot upside-down on the tray.
2. Through the holes in the bottom of the pot, puncture the tape and/or clingfilm. This lets the air in, so that the jelly can come away from the bottom of the pot.

3. Carefully lift off the pot to reveal your jelly tower. If the jelly sticks, try running a little tap-hot water over the outside of the pot to melt a thin layer of jelly.

4. With the help of a friend, quickly take a photograph of the tower with a tape measure beside it to prove its height. In our experience even good reinforced towers gradually slump; so take the picture as soon as you can.

LIFT OFF THE POT

5. If you do succeed in making a jelly tower 30cm high or more, please send us a photograph to prove it. We would like to know how high a jelly tower can be built, and what is the best technique.

More things to try

● Making a full-scale tower is expensive and slow. You might like to devise a way of doing the following tests on a small scale. For example, you might be able to make practice jellies in a 10cm flowerpot, and see how big a weight you could put on top of each one before they collapse. These tests would be a useful way of finding the best conditions for making a big jelly tower.
● Try making the jelly double-strength. Does it help?
● Try various different sorts of pasta, such as spaghetti and macaroni. Which do you reckon gives the best results?
● The water from the jelly softens the pasta. Can you devise a way to prevent this from happening? Or do you think it is necessary for the jelly to bind to the pasta?

WHY IT WORKS... Jelly is a poor constructional material for two reasons: it deforms easily under pressure, and it also has very low tensile strength – pull it and it breaks easily. Cement is surprisingly similar; mix it with water and it sets very hard, but not very strong, and in particular it is weak in tension.

As we said earlier, pasta holds jelly together just as aggregates hold cement together. Although pasta-reinforced jelly still deforms under its own weight, the pasta greatly increases the tensile strength; so reinforced jelly does not easily break when it is pulled. When something breaks, the break usually begins at a weak point in the material. By putting aggregate into cement or jelly you greatly reduce the length of unreinforced portions, and so reduce the chance of there being a weak point.

SPEAKING MACHINE

When Erasmus Darwin built his speaking machine in 1771, he did so to impress the assembled genius of the Lunar Society. All we know about his machine comes from an account in one of his books, _The Temple of Nature_. This and expert advice from professors of phonetics was all I had to go on.

RATING: ✳✳✳✳ DIFFICULT

The key to getting any kind of recognizable sound out of this machine is (perhaps not surprisingly) to mimic the human vocal cords and various passages as closely as possible. My machine was never as good as Erasmus Darwin's, but it definitely could say the vowel 'a' and the consonant 'm'. However, unlike Darwin's machine, the 'p' and 'b' were a bit dodgy.

You will need
- Approx 60cm of 32mm plastic pipe – the sort used for waste fittings
- A tee joint to fit the above piping (a swept tee)
- A right-angle bend to fit the above piping (a knuckle bend)
- Some plywood or MDF or any few bits of wood you may have hanging about
- Glue and sticky tape
- 2 blocks of balsa wood, 2.5 x 2.5 x 5cm
- A foot pump (the kind used to inflate your lilo)
- A length of ribbon, about 5mm wide
- A bamboo cane or piece of dowel
- A few elastic bands
- Nails
- 10 x 5cm sheet of plastic, 1mm thick

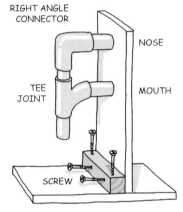

RIGHT ANGLE CONNECTOR

NOSE

MOUTH

TEE JOINT

SCREW

WOODEN MOUNT

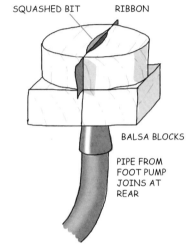

SQUASHED BIT   RIBBON

BALSA BLOCKS

PIPE FROM FOOT PUMP JOINS AT REAR

What to do
1. Start by making an F-shaped arrangement with the pipe and connectors. The distance from mouth to vocal cord (at the bottom of the 32mm pipe) should be 17cm, or a bit more. Keep the distance between nose and mouth as small as possible (but you should be able to see the pipe between the connectors).

2. Mount your contraption on a stand made from your assortment of wood or MDF. Cut two holes for the mouth and nose and fix the pipes in place with glue. Note that the mouth and nose should sit flush with the front surface of the wooden mount.

3. To mimic the vocal cord, the length of ribbon will be placed so that it vibrates in a flow of air. Take the two blocks of balsa wood and tape them together to create a block 2.5 x 5 x 5cm. Through the middle of one 5 x 5 side (where the blocks meet) drill a hole to take the foot-pump nozzle.

Drill to only half the depth of the block (1.25cm). On the opposite side of the block, make a flying-saucer-shaped hole as shown, that connects with the first hole. Do this by un-taping the blocks and squashing the edge of the balsa. The ribbon will lie in this hole.

4. Now, using a craft knife, carve the side of the balsa blocks with the squashed bit so that it will wedge into a piece of 32mm pipe. (It will probably help you to make the circular shape if you first press the end of the pipe against the balsa wood to make a mark.) Thread the ribbon so that it lies taut across the flying-saucer-shaped hole.

5. Finally, tape around the remaining square edge of the block. The end product should let you blow gently through the hole, across the ribbon, causing it to vibrate gently – thus making a buzzing sound.

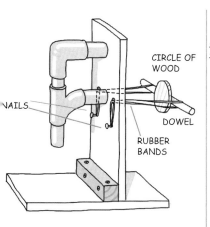

NAILS

CIRCLE OF WOOD

DOWEL

RUBBER BANDS

DIAGRAM 1

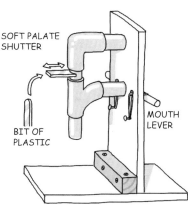

SOFT PALATE SHUTTER

BIT OF PLASTIC

MOUTH LEVER

DIAGRAM 2

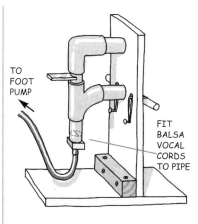

TO FOOT PUMP

FIT BALSA VOCAL CORDS TO PIPE

DIAGRAM 3

6. To make the mouthpiece, use a circle of wood, with a dowel or bit of bamboo stuck to the back to make a handle (diagram 1). To hold it tightly in place, drill two holes on either side of the mouth hole. Thread elastic bands through the holes, around the dowel, and fix the other ends using a couple of nails. You should now be able to use the dowel to lever open the mouth. The rubber bands will snap it back into the closed position again.

7. The second critical control is the soft palate (diagram 2). Cut a slot halfway through the upright pipe, between the nose and mouth. Then cut a piece of the plastic sheet so that it can be slid into the slot, effectively shutting the nose off from the mouth.

Time to speak

1. To make your machine speak you first need to practise with your own voice. Try sounding out the letter 'a'; as in the word 'tar'. Your tongue is lying flat in your mouth and air is coming mostly out of your mouth (if you pinch your nose the sound barely changes). If you close your mouth (while still making the 'a' sound – and not pinching your nose), the noise miraculously changes into an 'm' as in mother. Darwin's machine should be able to make both these sounds.

2. Now to use your speaking machine (diagram 3). Attach the vocal cord and foot pump – practise gently pressing the pump, so that you get a mournful buzzing from the mouth. Now, if you shut the soft palate (push the sheet into the slot) and open the mouth, you get 'a' (or nearly). Shut the mouth and open the soft palate, and you get 'm' (or thereabouts).

It took me quite a while to get the technique just right – and trying to combine the 'm' and 'a' sound into 'mama' required a fair amount of practice (clamping the stand to a table helps). As for making 'p's and 'b's – I have a feeling my mouth would need to be redesigned. Then there is always the possibility of a tongue, which in theory would mean you could sound the whole alphabet.

ADAM SAYS

If you manage to get any other letters out of the machine, you've done better than me. Maybe you should even try to fulfil Darwin's bet. His great mate, Matthew Boulton, offered Erasmus £1000 – a huge sum of money at the time – for a machine 'capable of pronouncing the Lord's Prayer, the Creed and the Ten Commandments in the vulgar tongue'. There are no reports of Boulton ever feeling nervous about the fate of his money.

TEST YOUR STRENGTH

In the 19th century, David Kirkaldy built a magnificent machine to break things. He wasn't a particularly destructive chap – he just realized that no one knew how strong things really were. This simpler machine can test the strength of things in three different ways.

You will need

- 2 pieces of U-section plastic track (see below)
- Glue
- 5 lengths of 25 x 50mm timber, each 15cm long
- Chipboard baseboard
- A piece of chipboard or MDF, 150 x 75mm, to slide in the U-section
- Assorted timber off-cuts
- Empty baked bean can
- Metal brackets and bolts (optional)
- A screw-eye
- Stiff wire (coat-hanger will do)
- Bottle tops or plumbing fixtures for specimen mounting (see below)

The plastic track is used for sliding doors to run in – plastic electrical conduit would do as well. The gap in the 'U' should be wide enough to allow the 150 x 75mm section to slide freely inside it.

What to do
The compression-testing machine

Make this machine first, as the others are easy modifications of it.

1. Start by making the uprights. Glue the lengths of plastic U-section track to the wider face of the 25 x 50mm timber lengths. When they have set, place them roughly in position on the baseboard, with the 150 x 75mm piece of chipboard or MDF between the U-sections, so that it can slide up and down freely but not fall out. Mark the position of the uprights.

2. Fix the uprights to the baseboard. Screw from beneath, perhaps with a touch of glue in the join. It is important that they are both upright and parallel, because the weight will slide between them.

3. Make a 'saddle' to sit on top of the sliding chipboard. Use three bits of wood to make a mount for the tin can, and glue and nail it to the top of the chipboard. Fix the tin can to the saddle. Drill a hole through the centre of the can, and screw it to the saddle.

4. Now you can test specimens for their resistance to compression. Wedge a specimen (Italian breadstick is good for a test) under the sliding chipboard while the can is filled with gravel, ball-bearings, marbles or another dense 'fluid'. We used lead shot but, unless you're a particular

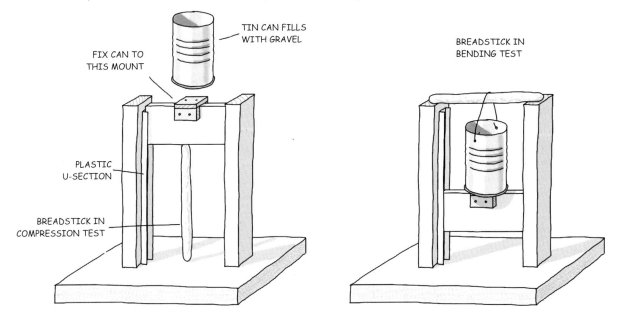

TIN CAN FILLS WITH GRAVEL

FIX CAN TO THIS MOUNT

PLASTIC U-SECTION

BREADSTICK IN COMPRESSION TEST

BREADSTICK IN BENDING TEST

sort of sportsman, you won't find this easy to come across.

Note: You may need to scale up the size of the can. Gravel is a lot less dense than either ball-bearings or lead shot, and you will need more of it.

THE BENDING-TESTING MACHINE

Modify the compression machine by drilling two holes on either side of the rim of the can. Make a wire loop, which will be used to bend the specimen. This time, the can is under the specimen, but is again filled with gravel to exert a load.

THE TENSION-TESTING MACHINE

This is the most tricky to make.

What to do

1. You need to fit an extension gantry over the original machine. Make the U-shaped gantry with your remaining pieces of wood, so that it is the same width as the original compression machine.
2. Attach the gantry with straight metal brackets and bolts, or you could make a bridging piece from some leftover wood, and screw it. Fix a screw-eye underneath the centre of the gantry.
3. The tension test is the most challenging because it is so hard to fix the specimens without affecting their strength. If you were using a breadstick, it makes a good specimen for the compression and bending tests, but you couldn't drill a hole through it to thread a bit of wire through, or grip it in any way for the

BREADSTICK IN TENSION TEST

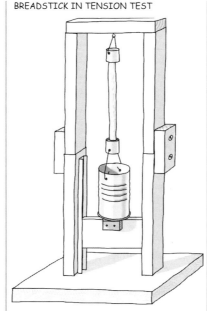

'stretching' or tension test.
4. For specimens such as breadstick, we mounted them in little bits of plumbing pipe, set in epoxy resin (something like Araldite). We drilled the pipe, which is able to take it, and mounted the breadstick in that.

An alternative approach is to make a specially shaped 'test piece'. When testing say, carrot, we would carve the centre section of the carrot to just 1cm diameter. This was to be the test portion. The ends were left full size, for mounting, and could be drilled without becoming too weak.

More things to try

● If you're going to test the strength of materials, you need to have a standard test. Now you have made a

nice testing machine, you just need easily compared specimens.
● There are two approaches here, both adopted by Kirkaldy. The first is to test real objects – such as girders and rails – to see what sorts of loads make them break. He tested the strength of various bits of Blackfriars Bridge, and was called upon to test the failed bits of the collapsed Tay rail bridge.
● His second aim, and the one your machine is best suited to, is to find out what material is best for what sort of job. You could settle on, say, specimens of 1 x 10cm.
● What should you test? Well, the machine is only made of MDF or chipboard so it isn't going to break any steel. To get you started, we suggest:
  ● Breadsticks
  ● Biscuits – which brand is strongest?
  ● Carrots
  ● Celery
  ● Chicken bones
  ● Spaghetti – and other pasta shapes
  ● Drinking straws

How to measure your result

● This is a simple test, and we want a straightforward result: what load breaks the specimen? You will have to decide whether, say, a 1cm-diameter breadstick is comparable to a 1mm-diameter piece of spaghetti. They are if they are being compared for the same job – a suspension bridge link.
● To measure the result, simply measure the depth of gravel (or whatever) in the tin when the specimen breaks.

DUELLING MACHINES

Another way to find out which of two materials 'wins' a particular test is to hold a race. First you will need to make a second, identical set of testing machines. Then make this automatic weight dispenser.

You will need
- 1 large tin can or plastic tub. It must hold as much as both tin cans from the testing machine together
- 2 lengths of U-section plastic track (see below), each about 30cm long
- 4 short lengths of stiff wire, each about 5cm
- 2 strips of wood, 1cm x 3mm x a bit taller than the tub (see below)
- Lead shot, gravel, ball-bearings etc for the flowing 'weight' (see below)
- Length of 75 x 75mm fence post to act as support

Note on materials
You are going to use this dispenser to allow lead shot (or similar heavy material that flows) to simultaneously fill the two cans of two side-by-side testing machines. Lead shot is difficult to obtain and a bit nasty – lead can be poisonous if swallowed. So you may choose fine gravel, cat litter for example, or small marbles instead. Make sure that you have enough room in your can for 2kg (you need 1kg for each machine). You may also need to adjust the size of the holes you make and the U-section you use. The strips of wood should be wide enough to cover the holes you make, and small enough to fit into the U-section.

What to do
1. Make two holes about 1cm square in the side of the tub, right at the bottom and separated by about 5cm. These need to be large enough for your lead shot or other heavy 'fluid' to flow through smoothly.
2. Cut about 3cm off the sides from one end of each piece of U-section, leaving just the bottom.
3. Fix the U-sections to the tub by gluing the 3cm flap you have just made to the bottom of the can, directly below the holes. The lead shot will pour out of the holes and down the U-section.
4. When the glue is dry, bend the track down slightly.
5. Now make the gates. These are the strips of wood that pass over the front of the holes. Fix two pieces of wire to make a 'guide' that will hold the gates in place, but which will still allow you to slide them up and down

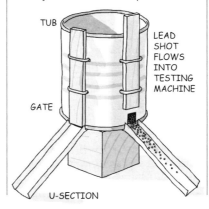

TUB

LEAD SHOT FLOWS INTO TESTING MACHINE

GATE

U-SECTION

fairly easily to start the race.
6. Finally, put everything together. Put your two identical testing machines side by side. Work out where you have to hold the dispenser so that the lead shot or gravel will flow into the tin cans of the testing machiens, and then cut a length of the 75mm fence post for the tub to stand on. You may need different lengths for each of the compression, tension and bending machines.

Running the race
Set up the testing machines with a different material, ie breadstick or spaghetti strand, in each. Slide the gates into place on your weight dispenser. Fill the tub with lead shot or whatever you are using. When you are ready, raise the gates simultaneously. One of the materials you are testing will fail first. As it does so, ram the gate shut immediately. Do the same for the second machine. This is a great way of seeing which of two materials is the strongest, and if you shut the gates when the samples failed, you can find out afterwards at what weight they broke.

WHY SHOULD WE CARE?
On 28 December 1879, the Tay Bridge collapsed as a train crossed it, killing 75 people. It was Britain's worst engineering disaster. When Kirkaldy investigated, it was found that the designer had wrongly estimated the stresses the bridge would be under in the severe gale blowing that night.

HOT AND COLD WATER

The Elizabethan philosopher Francis Bacon recorded a curious phenomenon; he said that hot water freezes more quickly than cold water. He didn't claim this was his discovery; indeed, he said it had been known for a long time. But is it true?

You will need
● A freezer
● 2 identical containers – glasses, mugs, tin cans or ice-cube trays

What to do
1. Fill one container with cold water from the cold tap and an identical one with hot water from the hot tap. Put both in the freezer, side by side. Make sure neither is touching a wall or anything else.
2. Check the two containers every half-hour or so. The aim is to find out which is frozen solid first. This is not always easy to tell, once both have a 'lid' of ice. However, often there will be a bit of water within the ice, and a bubble of air in the water. When you tip the container, you can see the bubble bobbing about in the liquid; then you know the water is not all frozen.

More things to try
● Try different containers. If the crucial factor is contact with the cold surfaces in the freezer, then tin cans should give better contact and so increase the effect. Take two identical tin cans – e.g. old baked bean tins – and try the experiment in them.
● The shape of the containers may provide a clue as to what is going on. For example, if cooling by evaporation is important, then a large surface area will help. Try the experiment with wide, low containers – margarine tubs or sardine tins.
● Try standing both containers on a metal surface (e.g. a roasting tin) in the freezer, to ensure equal contact – or on an insulating surface such as a plastic or ceramic plate.
● Try insulating your containers, by standing them on a layer of floppy polystyrene or newspaper and wrapping them in the same material. What happens then?

● Try using boiled water, which has little dissolved air. Boil some water in a kettle or pan. Let it cool. Put some in one container. Warm some of the water up again and put in the other container. Then carry on as before. Try using very hot water – just boiled.

BEWARE: DO NOT SCALD YOURSELF. DO NOT PUT BOILING WATER INTO ORDINARY GLASSES. DO NOT MELT ANY FOOD IN THE FREEZER.

Do write and let us know what results you get.

THE SCIENTIFIC METHOD
Francis Bacon was scathing about Aristotle's ideas that the truth can be discovered by argument alone. Bacon was one of the first people to advocate experiments, and summed up his idea neatly: 'Whether or no anything can be known, can be settled not by arguing, but by trying'.

WHY IT WORKS... (if it does!) Why hot water might freeze first is a bit of a mystery. One possibility is that if your freezer is frosted up, the warm container melts through the frost and gets better contact – so it then cools faster. Another is that a thin lid of ice forms quickly on the cold water, which actually helps to keep some of the heat in by stopping convection currents in the water. A third possibility is that bubbles of air inhibit freezing, and cold water tends to have more dissolved air than hot water, which might slow its freezing.

One simplistic explanation depends on the idea that the hot water starts to cool more quickly and simply goes on cooling more quickly, because heat has somehow got into the habit of leaving the hot container fast. This does not really make sense, because there must come a time when the water which started hot (say 60° C) reaches the temperature at which the cold water started (say 20° C). From then on it must surely cool at the same rate as the cold water did – but it has already taken some time to get there, and therefore it must always be lagging behind.

WHIRLING-ARM MACHINE

When the Wright Brothers flew their first aircraft in 1903 they paid a handsome tribute to the man who had started the science of aeronautics more than a hundred years earlier – a Yorkshireman called Sir George Cayley. He did the first whirling-arm experiments in the 1790s, and now you can make your own machine and use it to do real research.

You will need

- Screws (or nails and glue)
- A piece of wood about 30 x 30cm, for the upright
- A piece of wood about 60 x 30cm, for the base
- A nail or a ballpoint from an old pen, and an old spoon to rest it in, for the bearing
- A piece of 2.5cm dowel (or broomstick) about 60cm long (or 35mm plastic pipe), for the shaft
- 2 large rings or hooks, or an old coat-hanger, for the guides
- A piece of bamboo or dowel about 15mm in diameter and 70–90cm long, for the arm
- A piece of envelope stiffener about 30 x 30cm, for the wing
- A piece of tubing 15–20cm long and 15mm diameter (e.g. the centre from a roll of fax paper, or a length of garden hose)
- A nail for a spindle
- Bolts or other weight to balance the wing
- A pulley wheel
- About 3m of string, strong enough to carry 500g
- A weight of about 500g

What to do

1. Screw – or glue and nail – the upright to the base, with one edge of the upright at the centre of the base and one at the centre of one edge.

2. Staple the handle of the old spoon to the base, with the bowl next to the edge of the upright.

3. Cut a slot in the top end of the shaft, wide enough for the arm to be loose in it.

4. Fix the bearing to the bottom end of the shaft. I used plastic pipe for my shaft, inserted a plug of softwood, drilled a small hole, and pushed in the metal tip of an old ballpoint pen, so that the shaft rests on the ball point in the bowl of the spoon. Drilling a hole and then pushing a nail in backwards would probably be easier.

5. Fix your guides into the edge of the upright. I used large screw hooks, but you might prefer just to make these from wire (e.g. an old coat-hanger).

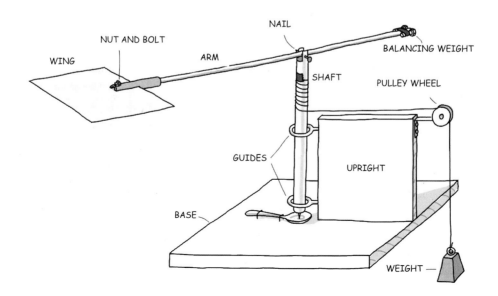

6. Make the arm. I cut a twig from a plant in the garden, but bamboo or dowel would be good. Near the centre drill a small (say 2mm) hole through it (if you're using bamboo, drill the hole in one of the lumps, which will lessen the chance of splitting).

7. To fix the wing to the arm you could just use tape, but it's important to be able to vary the angle of the wing. I used the centre of an old fax roll, which was a push fit on the end of the arm, and I made this comfortably tight with a piece of tape on the arm. Then I cut a 3mm slot in the end of the fax-roll tube, slipped the wing into it, drilled a hole, and bolted it through with a small nut and bolt.

8. Drill holes in the sides of the slot in the top of the shaft, and use a nail as the spindle for your arm, pushing it through the holes in the sides of the shaft and the arm. The arm should be free to tilt up and down.

9. With the wing in place, balance the arm by taping some weight to the other end. I used two coach bolts, but nails or Blu-Tack would do.

10. Fix the pulley wheel to the top outside end of the upright – mine just screws in.

11. Tape one end of the string to the shaft near the top and wind most of the rest around the shaft. Take the other end of the string over the pulley and tie on the weight.

12. Put your whirling-arm machine on the edge of a table, and it's ready to use. It's a scientific instrument, and you can decide what to investigate. Sir George Cayley was particularly interested in the best angle of attack – that is, he wanted to find out what angle of the wing would give the best combination of lift and drag.

13. To do this, set the wing horizontal. Make sure the arm is balanced and stays horizontal after you have adjusted the wing. Release the weight, so that the string makes the shaft rotate and the arm whirls.

14. Note whether the wing rises or falls (it should do neither) and how fast it goes round (e.g. how many revolutions it makes in 10 seconds).

15. Wind up the weight again. Twist the wing so that as it whirls round it makes an upward angle of, say, 45 degrees to the horizontal.

16. Release the weight. Does the wing rise? If so, you have lift – an upwards force on the wing. How much more slowly does the arm whirl? The whirling is slowed because in addition to lift you also have drag – a backwards force on the wing, like friction, and commonly called air resistance.

17. Vary the angle between the wing and the horizontal – the angle of attack. Find out what angle gives you the optimum combination of lift and drag. Cayley found that for his whirling-arm machine the best angle was about 6 degrees.

More things to try

● Plot a graph of the whirling speed (number of revolutions in 10 seconds) against the angle of attack. Does the drag get steadily worse, or is there a sudden change at some specific angle?

● Try different air-speeds, by varying the weight on the end of the string. Does this make any difference to the optimal angle?

● Be more scientific by actually measuring the lift. Hang small weights such as paperclips on the outer end of the wing. For each angle of attack, see how many you can hang so that the arm just gets back to horizontal when whirling. Write your results down, or you will lose track!

● Plot a graph of the lift (number of paperclips) against the angle of attack. Looking at this graph and the previous one, what angle of attack would you choose if you were going to design an aircraft?

● Try different wings, with various sizes and shapes. What effect do size and shape have on the lift?

DID YOU KNOW?

In 1853 Sir George Cayley achieved the world's first flight by a man in a heavier-than-air machine. He was 79, and so he volunteered his coachman John Appleby to be the world's first test pilot. Appleby flew about 200 metres across Brompton Vale, crawled from the wreckage after a heavy landing, and said, 'Sir George, I wish to resign. I was hired to drive – not to fly!'

Cayley did his whirling-arm experiments in the magnificent stairwell of his home, Brompton Hall, near Scarborough. His wife disapproved, so he waited to set up the apparatus until she went off to stay with her mum to have their first baby!

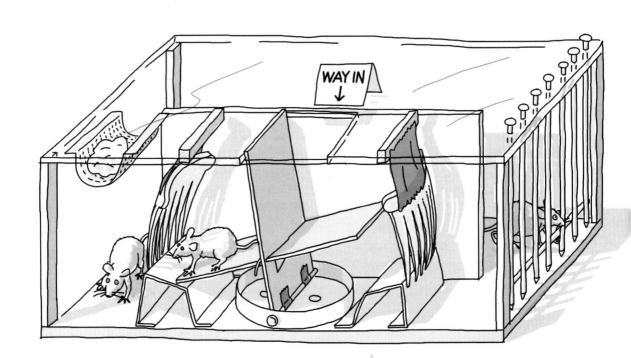

CHAPTER NINE
INSTRUMENTS

THERMOMETERS

Before the reliable thermometers made and calibrated by Daniel Fahrenheit in the early 18th century, there was no universal way of measuring temperature. Here are two primitive thermometers, as used by Galileo in Italy and Denis Papin in London.

GALILEO'S AIR THERMOMETER

RATING: ✳ EASY!

You will need
- A plastic canister for 35mm film, with lid
- The thin inner tube of a used ball-point pen
- Vaseline
- Food colouring
- Waterproof marker pen

What to do
1. Make a hole in the lid of the film canister so that the tube from the ballpoint pen fits through very tightly, but without squashing the tube flat. Seal round the tube with a blob of Vaseline.
2. Fill the canister a third to half full of water, and add a drop or two of food colouring so that you can see it.
3. Put a little Vaseline round the lid of the canister to help seal it, then fit the lid back on. Push the tube down until it is just above the bottom of the canister.

Calibrating your thermometer
If you grip the thermometer in your hand, the air in the canister expands and the water level will rise up the tube. Mark the level on the tube with your pen – this is 'hand hot'. Put the thermometer into cold, warm or hot water. Make marks at useful temperatures, such as 'hot tea' or 'gin and tonic'. Unfortunately this thermometer is not very good at boiling point or above, because the water in the thermometer itself will boil, and neither is it very good at

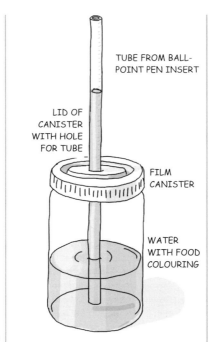

TUBE FROM BALL-POINT PEN INSERT

LID OF CANISTER WITH HOLE FOR TUBE

FILM CANISTER

WATER WITH FOOD COLOURING

freezing temperatures, because, of course, it freezes below the freezing point of water.

Nevertheless, this is how Daniel Fahrenheit started his temperature scale. He chose fixed points like freezing water, boiling water, and 'human armpit' to be 32, 212 and 96 degrees respectively. Freezing water was 32, to leave room for a mixture of water and ice, the coldest thing then known, to be zero. So you can see that Fahrenheit's scale of temperature doesn't make much more sense than yours! But it was universal, and meant that for the first time scientists could discuss heat and cold knowing exactly what was meant.

ADAM SAYS
Galileo's thermometer is not very accurate, not least because it is greatly affected by air pressure. Fahrenheit realized this, and actually constructed a barometer based on the same design! Fahrenheit switched to using sealed mercury in glass thermometers where air pressure at least is not a problem.

DENIS PAPIN'S PENDULUM THERMOMETER

RATING: ✳ EASY!

Papin worked at the end of the 17th century, and had neither a decent thermometer nor clock. When he invented the pressure cooker (he called it an 'engine for softening bones') he needed some way to tell how hot it had got. This is his method. See if you can use it to tell when water is boiling in a saucepan without taking the lid off.

You will need
- A piece of string, about 1m long
- Blob of Plasticine or similar to act as a pendulum bob
- Thin tube (plastic tube used above would do)

The tube is for making fairly regular water drops. You could use a stick dipped into water.

What to do
1. Make a pendulum by fixing the blob of Plasticine or similar weight to the end of the piece of string. Each swing of the pendulum, which you could hold but would be better tied to the back of a chair, takes about 1 second.

2. When you are cooking, put a drop of water on to the lid of your saucepan. Immediately set the pendulum swinging. How many swings of the pendulum does it take for the drop to evaporate? As the saucepan heats up, you can add drops to measure the increasing temperature. You could calibrate this with a 'real' thermometer, but make sure it is one designed to go in boiling water.

3. You should see a dramatic reduction in the time the drop takes to disappear when the water boils. Papin expressed his temperature results as: 'I kindled the fire until a water drop evaporated in twelve seconds.' It doesn't mean anything except to anyone repeating the exact same experiment, but the clever thing is that if you did repeat what Papin had done, even without a thermometer you could guarantee to get exactly the same temperature.

DENIS PAPIN AND THE PRESSURE COOKER

Why was Denis Papin bothering to measure the temperature of his pressure cooker so very carefully? Papin's name for his new machine – an 'engine for softening bones' – explains his purpose. We use pressure cookers mainly for reducing the cooking time of stewed foods. Normally, water boils at 100° C, and no matter how much heat you put in, it doesn't get any hotter. But as you increase the pressure, the boiling point goes up as well. Papin had probably discovered this while working for Robert Boyle, whose 'Boyle's Law' sets out the relationship between temperature, volume and pressure in gases.

Papin seems primarily to have used his engine for making edible foods that normally would not have been fit for the plate. So, for example, in experiment 11, he took an 'old male and tame Rabbet [sic], which is ordinarily but a pitiful sort of meat'. He kindled six ounces of coal until the water drop evaporated in four seconds, and the pressure in his digester was about six times stronger, he said, than normal air pressure.

The result was very good: 'the Rabbet was well cooked, the bones softened, and it tasted as good as young Rabbet.' The bones seem to have been crucial. Papin was interested in producing different sorts of 'Gelly' from bones – we guess it was a sort of stock thickened with gelatine and fat. This would make very tasty and very cheap food for the poor, he thought.

Papin's experiments were very carefully done. In those days, controlled experiments were not all that common. But Papin, with his careful measurements of the exact conditions of each firing of the pressure cooker, the ingredients, the temperature and pressure attained, provided a blueprint of how these things should be done.

As with many heroes, Papin did not receive the credit he deserved during his lifetime. In particular, his experiments on the steam engine are not as well known as those of Savery, Newcomen or Watt, even though Papin beat all of them to some aspects of steam technology. In this case, Papin's downfall was partly of his own making. He believed in sharing information, and rushed his results into print before they got perfected – often leaving others to pick up the baton and get the credit!

BAD BAROMETERS

Why would we encourage you to make anything that wasn't very good? In this case, because to make a real barometer – a device to measure air pressure – is extremely difficult. These two barometers do measure air pressure, but only if you take great care. So maybe they aren't all that bad; we could have called this 'barometers for centrally heated homes'.

TIN CAN BAROMETER

RATING: ✳ EASY!

This barometer is actually an aneroid one (i.e. 'without water'). Most of the barometers people have in their hallways (and lovingly tap) are aneroid.

You will need
- Empty tin can – standard baked bean size or larger
- Sticky tape
- Large balloon
- Card
- Drinking straw (not bendy), the longer the better
- Glue

What to do
1. Wash and dry the can, and cover the sharp edge with sticky tape – this is to protect you (and the balloon). Cut the neck off the balloon, leaving you with a sort of bathing-cap shape. Stretch this over the can, and then seal all round the edge with tape. The idea is to get it airtight. You should try to trap a little extra air in the can, so that the balloon just sticks up above the rim, rather than down. This is not critical, but it is best not to have the balloon tight.
2. Now make the scale. Cut a T-shaped piece of card with the upright about 3cm across and a little taller than the can. Fold the T down the centre of the upright and glue the upright where you have folded it. Leave the cross-pieces of the T unglued.
3. Turn the T upside-down, open out the cross-pieces, and fix to the can as shown using sticky tape.

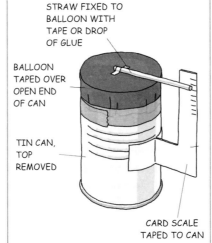

STRAW FIXED TO BALLOON WITH TAPE OR DROP OF GLUE

BALLOON TAPED OVER OPEN END OF CAN

TIN CAN, TOP REMOVED

CARD SCALE TAPED TO CAN

4. Finally, the straw. Fix one end to the centre of the balloon using either a tiny strip of tape, or a little blob of glue. Arrange the straw so that it also rests on the rim of the can and passes just in front of the scale.

Using your aneroid barometer
The air you sealed inside the tin can is at atmospheric pressure – about 14.6 pounds per square inch, or 101 kP (kilo Pascals) or Newtons per square metre at sea level, and a temperature of 20 degrees C. But don't worry: the balloon isn't going to burst with this huge force (though there really is that force pressing on the inside) because there is an exactly equal force pressing on the top of the balloon.

The idea, of course, is that if the air pressure outside the can changes, it will press down a bit more or a bit less on the balloon. This will push the balloon down a bit, or let it spring up a bit. The balloon in turn is connected to the straw, which will go down or up the scale.

So the first thing to do is to mark the position of the straw on the scale. Look at it over the next few days and weeks, to see if it moves up or down.

The next obvious thing to do is to calibrate the scale. You could do this by getting the air pressure (sometimes called barometric pressure) from the weather report in your local newspaper. However, this is where we come up against the limitations of this particular barometer. All would be fine if air pressure were the only thing that makes the balloon go up or down. But it isn't. Higher temperatures will make the air in the can expand, lower temperatures will make it contract, and it's impossible to tell whether the temperature or the pressure is making the straw move in any case. Indeed, these sorts of devices are sometimes known as thermo-barometers.

So is it useless? No – on two counts. First, you might be able to relate the movement of the straw to the weather in any case: perhaps a combination of temperature and pressure will be as predictive as pressure alone. Second, the reason we suggested this might be called a 'barometer for centrally heated homes'. If you keep the barometer at the same temperature all the time, then only air pressure will affect it, and modern homes where the heating is controlled by a thermostat have fairly even temperatures. You could even check the temperature by keeping a thermometer near to your barometer. Once you have taken this precaution, your machine becomes a barometer again, and you can indeed calibrate it by marking the 'known' pressures from the newspaper weather report.

In fact all barometers have to take account of temperature if they are to be really accurate, because the glass and metal they are made of also expand and contract to some extent with changing temperature.

'BAD' WATER BAROMETER

RATING: ✳ EASY!

You will need
● Chewing gum
● Water with food colouring in it
● A glass beaker
● A clean, clear tube – a drinking straw, a used ballpoint pen insert, or ballpoint pen barrel
● Waterproof pen for marking

What to do
1. Chew the gum. Fill the glass about half-full with coloured water. With the gum still in your mouth, put the tube into the liquid. Suck up liquid until it is between half and three-quarters of the way up the tube. Using your tongue, push the chewing gum over the end of the tube to seal it. Mark a scale on the tube with the waterproof pen.

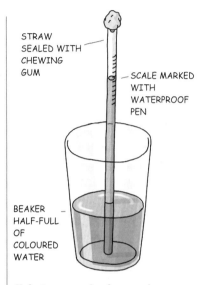

STRAW SEALED WITH CHEWING GUM

SCALE MARKED WITH WATERPROOF PEN

BEAKER HALF-FULL OF COLOURED WATER

Using your water barometer
This barometer works because you created a partial vacuum above the water when you sucked. There is slightly less air pressure above the water in the tube compared to that in the beaker – this must be true because the extra air pressure is holding the water up the tube.

So how far the water goes up the tube depends upon how much air pressure there is pressing down on the water in the glass. Once again, because you don't have a good vacuum in your tube, the air in the gap will also expand and contract with temperature. But, as with the previous barometer, this one will measure air pressure if you keep it at a constant temperature. The basic design of this barometer is the same as that of the best sort of mercury barometer. But even though that has

a vacuum at the top, which is not affected by temperature, all the other bits of a barometer do expand or contract with heat, so you still need to know what the temperature is.

TORRICELLI'S VACUUM
The barometer was invented by an Italian, Evangelista Torricelli, who was secretary to the great Galileo. Torricelli was the first person to show that a vacuum could exist – previously it was assumed that 'nature abhors a vacuum'. Torricelli put mercury into a tube that was sealed at one end. Putting his finger over the other end, he turned the tube upside-down (so the seal was then at the top) and put the open end into a beaker of mercury. If the tube was more than about 76cm long, a little mercury would run out – leaving a gap at the top. This gap at the top of mercury barometers is called Torricelli's vacuum.

MAKE A REAL WATER BAROMETER
The same physics applies to any liquid. So what about a real water barometer? After all, water is a lot safer than mercury, and easier to get hold of. Yes, but it is nothing like as dense. Because water is 13½ times less dense than mercury, you need a tube 13½ times as long to get a vacuum at the top. That is 10.2 metres or about 33 feet! If you try this, do not use a glass bottle that could collapse under the pressure of the atmosphere. But if you succeed, you'll have the most spectacular barometer on your street.

THUNDERSTORM PREDICTOR

The 18th-century English physicist John Canton was the first to detect and measure electrostatic charge – the stuff that makes your hair stand on end and balloons stick to the wall. Here are a couple of his charge-detecting devices for you to build.

THE ELECTROSCOPE

RATING: ✳✳ MODERATE

This was one of Canton's most important developments – before his electroscope, there was no way of comparing or measuring charge.

What you need

- Wood for upright, 2.5 x 2.5 x 25cm
- Wood for base, 15 x 15cm
- 15cm dowel or similar, 6mm diameter, for horizontal arm
- Screws
- Glue
- 2 polystyrene balls (about 25mm diameter)
- 30cm fishing line
- Wire, if necessary
- 2 pieces aluminium foil, about 10cm square

What to do

1. Fix the upright to the base by drilling a hole through the base, countersinking the hole so that the screw head won't scratch your table. Make a small pilot hole in the upright, and screw it to the base.

2. Now fix the arm to the upright. Drill a 6mm hole (if your dowel is 6mm diameter) and either simply shove the end of the dowel into the hole so it pokes slightly through to the other side, or glue it in for added strength.

Fixing the balls

1. Now string up the balls. Attach the fishing line to the two balls, perhaps by pushing a small piece of wire into each ball to form a loop and tying the line through it. Adjust the length of the line so that the balls hang clear when suspended over the arm.

2. Now cover both balls with the foil, tightly wrapping to cover the whole surface. Hang the balls up so that they hang at equal height, well clear of the base. Fix the line in place with a drop of glue.

Using the electroscope

The electroscope works by picking up electric charge on the foil coating on the balls. Because they both pick up the same charge, and like charges repel, the balls move apart. It will work better when the air is fairly dry, which helps to stop the charge leaking away.

1. Try rubbing a balloon or a piece of Perspex on your jumper. As you bring it up to the electroscope, the balls should move apart. But then try something else before the charge has gone. If the balls move further apart, then the new object has the same charge as the previous one. But if they move together, the charges are opposite.

BUILD A THUNDERSTORM PREDICTOR

RATING: ✳✳ MODERATE

When I made the thunderstorm detector it was a dull, but distinctly un-thundery, day in Stroud (Canton's home town). I used a Van der Graaff generator attached to an artificial metal cloud to demonstrate how it

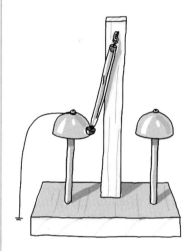

might work. When the programme went out, it was the night of the most severe thunderstorms of the year and several viewers got their device to ring without any assistance.

I guess physics teachers might have the odd Van der Graaff generator knocking about, but the rest of you will have to rely on other forms of static electricity like rubbed balloons or bits of Perspex. Don't use any other sort of electrical supply, and especially not the mains, though it should be all right to bring the detector up to the screen of your television – an excellent source of charge.

WARNING: THUNDERSTORMS CAN BE DANGEROUS, AND YOU MUSTN'T DO ANYTHING THAT COULD INCREASE YOUR RISK IF LIGHTNING STRIKES. DON'T MAKE AN AERIAL OR OTHER LIGHTNING CONDUCTOR, AND DON'T CONNECT SUCH AN AERIAL TO THE EARTH OF YOUR HOME, SCHOOL OR WORK ELECTRICITY SUPPLY.

What you need
- Wood for upright, 30 x 2.5 x 2.5cm
- Wood for base, 15 x 15cm
- Screws
- 2 bicycle bell tops
- 2 dowels or similar, 12mm diameter, 15cm long, for bell supports
- Metal button, approx 1cm diameter (the sort you buy for covering with cloth is good, or a drawing pin might do)
- Plastic biro, ink tube removed
- Screw hook and eye for fixing pen casing
- Wire for earthing
- Small wood-screws for fixing bells

What to do
Make the base and bells
1. Fix the central upright to the base by drilling a hole through the base, countersinking the hole so that the screw head won't scratch your table. Make a small pilot hole in the upright, and screw it to the base.
2. Fix bicycle bell tops to the dowels using screws – it is a good idea to drill a pilot hole in the top of each dowel to avoid splitting wood.

3. Drill two holes part-way through the base for the dowels (10mm holes for 10mm dowel, etc), so that when the bells are in place their edges are about 2.5cm apart. Fix the dowels, with bells attached, relying on either friction or a little glue. It is important that the bells are equal distances from the central upright, since the ringer must hit both.

The ringer
1. Fix the metal button to the tip of the pen-casing, using glue, wire or a tiny screw-eye in the existing hole (where the pen insert used to be). The button we used had a nice wobbly fixing, so that it hung free. This probably is not important. Now fix a screw-eye into the plastic fitting in the top of the pen-case, and the hook into the central wooden upright, so that you can hang the pen centrally. It should swing un-impeded, and be able to hit both bells.

Earthing
1. One bell top needs to be earthed. You could use a water pipe, or a metal spike in the ground. DON'T FIDDLE WITH PLUGS AND SOCKETS AND, IF YOU'RE UNCERTAIN WHETHER WHAT YOU'RE DOING IS SAFE, PLEASE STOP!
2. Use the screw fixing on one bell to connect a length of conductor to the earth.

Using the detector
The idea is simple. Bring a charged object (or thundercloud) up to the un-earthed bell, and it will collect some of the charge. The bell will then attract the ringer because if the bell is, say, positive, the ringer (which is insulated) will be relatively negative. However, when the ringer touches the bell (making it ring) it will immediately collect the positive charge and be pushed towards the earthed bell. When it touches the earthed bell (making it ring) it loses its charge because it is conducted to earth. The ringer swings back, and the cycle starts again. You will get a continuous ringing sound.

DID YOU KNOW?
The safest place to be when lightning strikes is inside a car? The lightning travels easily through the metal skin of the car, leaving you untouched. Not much good for cyclists, though.

JOHN CANTON
John Canton was born in Stroud, Gloucestershire, the son of a non-conformist broad-cloth weaver. His only formal schooling was stopped when he was nine by his father, who decided that little John had received enough education to become a weaver like his dad.

But John would stay up late into the night reading and doing calculations. Eventually, the local gentry took notice and they let Canton use their libraries.

The best-known electrical experimenter was the American scientist Benjamin Franklin. The two knew each other and corresponded. Strangely, the thunderstorm predictor is usually known as 'Franklin's bells'. It's possible Franklin was doing the experiment at the same time. However, I am fairly sure Canton did it first and should get the credit.

PERPETUAL MOUSETRAP

Colin Pullinger was, among other things, a builder, baker, undertaker, fisherman, farmer, mender of glass, cooper, clock cleaner, collector of taxes, repairer of umbrellas and Clerk to the Selsey Sparrow Club! And, about 1860, he invented a new mousetrap.

A normal mousetrap kills a single mouse, and then won't work until it is reset. Pullinger called his mousetrap perpetual because it was always ready to catch mice; indeed he claimed that one of his traps caught twenty-eight mice in a single night!

What's more, a normal mousetrap is brutal. Pullinger's mousetrap is humane; the mice it catches are quite unharmed, and can be released from the trap in the country – or in your neighbour's garden!

You will need

- 1 sheet 9mm or 12mm ply (or MDF), 35cm x 12cm
- 80cm softwood, about 20mm x 15mm
- Glue, pins and screws
- 2, 3 or 4 sheets 4mm or 6mm ply or hardboard, 35cm x 12cm
- Optional replacements for 2 sheets of ply or hardboard: 1 or 2 sheets clear plastic – Perspex or acrylic – about 3mm, 35cm x 12cm. Using this stuff is a bit tricky, but it lets you see inside the trap.
- 1 sheet (A4 or C4) of flyweight plastic
- Old knitting needles or metal skewers, tent pegs or nails
- 1 old yogurt, cottage cheese or hummus pot
- 2 cheap plastic side combs (optional but neat)
- A piece of perforated zinc or expanded metal (or an old tea strainer) to hold the cheese bait

What to do
The basic structure

1. Make the base from the thicker piece of ply or MDF. Cut the barrier to size, 25cm x 10.5cm. Fix the barrier to the base, using 20cm of the softwood, with glue and pins, or angle brackets. The barrier should be vertical, and about 5cm from the back of the base.

2. Cut four corner posts, each 10.5cm long, from the softwood. Fix the posts to the corners of the base using glue and pins. If necessary trim your roof to the same size as the base, 35cm x 12cm.

3. Cut the entrance hole in the front edge of the roof about 5cm wide and 6cm from front to back. I marked this with a sign saying WAY IN for dim mice – and to explain the trap to human beings.

4. Screw the roof to the corner posts but do not glue it yet. You will have to remove the roof later for adjustments.

5. Cut the left-hand end from thin ply or hardboard to fit the base and corner posts, about 12cm square.

6. Fix it to the base and posts (but not to the top) with glue and pins.

7. Cut the back (35cm x 12cm) from either ply or hardboard or clear plastic. If you choose to use plastic (so that you can see what you have caught), carefully drill holes for fixing it to the base and posts.

8. Fix the back in place, using pins and glue if you have a ply or hardboard back, or screws for a plastic back.

9. Cut the front (35cm x 12cm) from plastic or ply, but don't fit it in place until the very end.

The rocker assembly

This is the vital moving part that enables the mousetrap to be perpetually set.

1. Cut a strip of flyweight plastic along the grain of the tubes about 5.5cm wide. Cut one piece 12cm long for the vertical part of the rocker, and another 18cm long for the horizontal part (15cm may be enough, if you have a piece that long).

2. Make these two pieces into a cross by cutting 2mm slots, half-way across the width, in the centre of the horizontal piece and 3cm up the vertical piece. Slot them together. Cut small triangular pieces and tape them on to keep the two pieces at right angles (see diagram).

3. For the rocking axle use an old cut-down knitting needle or long nail, ideally just 6cm long. Tape it temporarily along the bottom of the vertical rocker strip, with at least 1cm sticking out at the back.

4. For the bearings I used an old plastic yogurt pot with a flat bottom. Cut off the bottom 2cm, drill holes in opposite sides to fit the knitting needle loosely, glue and pin the pot bottom to the base in front of the barrier, and drill a hole in the barrier for the end of the knitting needle. You may prefer to use two screw-in eyes, or find a better solution.

5. With the axle resting in the bearings, the rocker assembly should poke out through the entrance hole, and rock from side to side of it. If you are satisfied that it fits well, trim off the top of the vertical strip until it is only just through the entrance. Your rocker assembly is now complete.

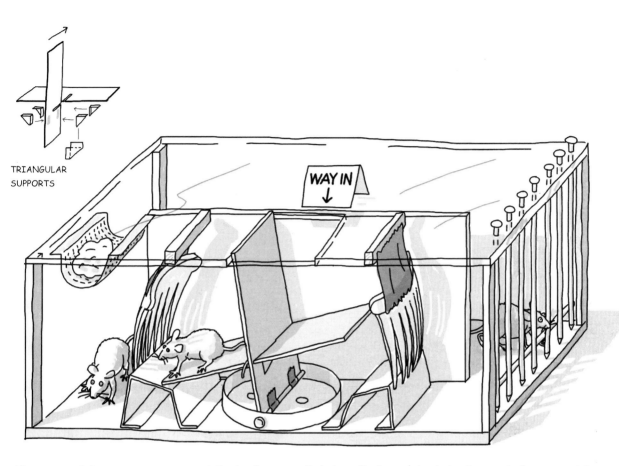

TRIANGULAR
SUPPORTS

WAY IN
↓

The mouse gates

1. Cut two 16cm lengths of flyweight plastic about 5.5cm wide, bend them into landing ramps, as shown in the diagram.

2. Tack the outer ends to the base first. Fix the inner ends temporarily with tape or clothes pegs.

3. Test the rocker assembly again. At each end of its travel the vertical strip should touch the side of the entrance, and the horizontal strip should touch the landing ramp, or nearly so. Adjust the shape of the ramps until this happens. When a mouse enters the trap, its weight on the horizontal strip tips the rocker assembly, which closes the entrance, and the mouse has to step off on to

the landing ramp. Tack down the inner ends of the landing ramps.

4. Cut two pieces of 20mm x 15mm softwood. Fit these pieces temporarily, using tape, to the underside of the roof, 5cm or 6cm from the ends, as shown.

5. Suspend the plastic hair combs from these pieces of wood using parcel tape so that they rest at an angle of about 70 degrees on the outer ends of the landing ramps. If you have no hair combs you can use pieces of plastic, 5.5cm square. Fix the pieces of wood.

6. The hair combs are the mouse gates; a mouse trapped on the landing ramp pushes through the mouse gate, but then cannot get

back. In other words the mouse gate is like a one-way valve. Each mouse that gets through a mouse gate is enclosed in the space at the ends of the trap and behind the barrier.

Finishing off

1. The portcullis at the right-hand end of the trap is for releasing the trapped mice. I made mine from knitting needles which slid through holes in the roof and into slots in the base, but you could use skewers, tent-pegs, or make a door from wood or plastic.

2. The bait rack is made from perforated zinc or an old tea strainer, and fixed to the roof in a spare space, so the smell of the bait fills the trap.

POCKET CALCULATOR

People have always found calculation difficult, and many devices have been invented to make the process simpler. Most people can add up without too much trouble, but multiplying is much harder. Napier's bones turn multiplication into addition. They were invented by Scottish astronomer and mathematician John Napier 400 years ago, and they still work!

You will need
- 1m square-section wooden rod, about 10 x 10mm
- Paper and glue, if necessary

What to do

1. Saw the rod into five 12cm lengths; keep the rest for spares. Glue the paper carefully round each section of rod. Note: you may leave this step out if you can write directly on the wood.

2. Use a ruler to draw a line across each side of each rod at each centimetre from the bottom. Note that you can draw across all five at the same time.

3. Use the ruler to draw diagonal lines from bottom left to top right of all the squares below the top one.

4. Mark the top square of the first rod 1 on one side, 2 on the next, 3 on the third, and 4 on the fourth.

5. Mark the top squares of the second rod 3, 4, 5 and 6.

6. Mark the top squares of the third rod 5, 6, 7 and 8.

7. Mark the top squares of the fourth rod 7, 8, 9 and 0.

8. Mark the top squares of the fifth rod 9, 0, 1 and 2.

9. Going back to the first rod, leave the first square below the 1 blank. Then mark the right-hand triangles 1, 2, 3, 4, 5, 6, 7, 8, 9 and 0, as shown in the diagram. Mark the left-hand triangle in the bottom square 1.

10. Still with the first rod, and below the top figure 2, again leave the next square blank. Then mark the squares

2, 4, 6, 8, 10, 12, 14, 16, 18 and 20, as shown in the diagram.

11. Follow this pattern with the other two sides of the first rod, and all four sides of the other rods: i.e. leave the second square under the top number blank; then mark each side with the times table (from 1 to 10 times) of the top number. So the side with 6 at the top will be labelled 6, 12, 18, 24 etc up to 60.

12. This takes a long time: to begin with, just mark up one side of each rod you have cut.

13. Now at last you're ready to use your calculator. Start with a rod with 1 at the top; this is the multiplying

rod (to multiply by numbers up to 10). Then you need the rods whose top numbers are the digits of the other figure in the multiplication sum. As an example, the diagram shows how to calculate 7 x 53. This is how to do it.

14. Find sides labelled 1 (the multiplying rod) and 5 and 3 (the digits in the other number). Lay these rods side by side, as shown in the diagram.

15. Because you want to multiply by 7, read across from the 7. The four numbers are 3, 5, 2, 1. The answer takes the first number, then the sum of the middle two numbers (which are together between the diagonal lines),

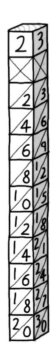

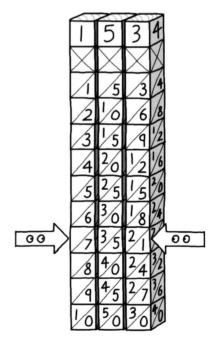

then the last number: 3(5+2)1, or 371. So 7 x 53 = 371.

16. To multiply 53 by 8, read across from the 8 – the answer is 4(0 + 2)4: 424.

17. To multiply 53 by 9, read across from the 9 – the answer is 4(5+2)7: 477.

More things to try

● Try multiplying by numbers greater than 9; for example, try multiplying 53 not by 7 but by 78. You will need to do this in two stages, and add the results of each stage together. These two stages are 53 x 70, and 53 x 8.
● First, multiply 53 by 7, to get 371 as before. Then write down the answer to 53 x 70, which is 3710. Next multiply 53 by 8, to get 424. Finally, add this to 3710 to get the answer 4134. Therefore 53 x 78 = 4134.
● Try using Napier's bones for division, which is tricky but possible. For example, to divide 3922 by 53, set up the rods as shown in the diagram – that is, set up the number you want to divide by (in this case, 53).
● Now look down the 53 rods for the three-digit number that is nearest to, and below, the first three digits of the number you are dividing into. (As before, the middle two numbers are added together.) The first three digits of your target are 392; the nearest number below this on the rods is 371, which is 53 x 7; so the first digit of your answer is 7, and your answer is going to be 70-something.

Since 7 x 53 = 371, 70 x 53 = 3710. Therefore, to find the rest of

the answer, subtract 3710 from 3922, to give 212, and divide 212 by 53. Again, look down the 53 numbers to find the nearest one below 212 – and you find that row 4 is 212. So the second digit you want is 4, and the answer to 3922 divided by 53 is: 74.

DID YOU KNOW?

John Napier was the eighth Laird of Merchiston, and had a castle, just outside Edinburgh. He invented logarithms and probably also the decimal point. He described the use of his rods in a book called *Rabdologia*, published in 1617. They were originally called Napier's rods,

but because posh sets were made from ivory, they came to be called Napier's bones.

Napier kept a black cockerel, which he claimed was psychic. On one occasion some valuables disappeared, and he suspected that one of the servants was a thief. He lined them up and told them to go one at a time into a dark room and stroke the cockerel, which he said would pick out the thief by crowing.

They all went in and came out, but the cockerel did not crow. But then he asked to see their hands. All but one of the servants had black hands because Napier had covered the cockerel with soot!

MAGNET & COMPASS

One day in 1818, when his whaling ship was stuck in the ice off Greenland, William Scoresby Junior made a magnet; he claimed it took him only 40 minutes. (So he had plenty of time to make fire from ice too – see page 88.) The magnet is still in the museum at Whitby, and it is powerful enough to pick up 4kg of iron. This is how to make your own Scoresby magnet.

RATING: ✱ EASY!

You will need
- A chunk of wood for a stand
- A compass (to test your magnet – otherwise use paper-clips)
- 3 or 4 pieces of iron, about 15cm long
- A solid base to hammer on – perhaps a large stone – which can be damaged without anyone minding
- A hammer
- Wood and a rubber band, if necessary
- String or tape

Six-inch (15cm) nails would do for the iron pieces, according to Scoresby, but our advice is to get strips of soft iron from a blacksmith if you can. Soft iron is much better than steel.

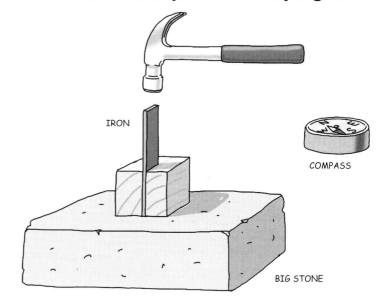

IRON

COMPASS

BIG STONE

What to do

1. Saw or drill a slot through your chunk of wood so that one of your pieces of iron will stand upright in it. You don't absolutely need this stand, but it does help to prevent you from bashing your fingers!

2. Use your compass to make sure the pieces of iron are not already magnetized. Iron attracts a compass needle, but if it is already magnetized then each end of the iron will attract one end of the compass needle and push the other away. If you do not have a compass, see whether any of your pieces of iron will pick up a paper-clip.

3. Take one piece of iron, set it upright in the stand on the solid base. Hit the top firmly with the hammer, about ten times, always making sure you hit the same end of the iron.

4. Now test the iron with the compass; does it repel one end of the compass needle? If it doesn't, then hit it ten times more, making sure you keep hitting the same end of the iron.

5. When the iron is slightly magnetized, set a new piece of iron in the base, and hold the first piece on top of it, end to end.

6. Now hit the top of the first piece with the hammer. Be careful not to hit your hand. If you are nervous, you could hold the upper iron with tongs, pliers, or a piece of wood and a rubber band, as shown in the diagram.

7. Again, hit the iron ten times. Both pieces of iron should now be magnetized more than the first piece was before you added the second piece.

8. Repeat this process with the other pieces of iron, remembering each time to keep the same end up. (Using nails is an advantage here; simply keep the head of each nail at the top every time.)

9. When you're exhausted or bored, bind all your pieces of iron together with string or tape, still keeping all the 'up' ends up. You should now have a fairly strong magnet, and if it can pick up a couple of paper-clips you have done well. If it can pick up a kilogram of iron, please write and tell us about it, and send a photograph of the magnet in action.

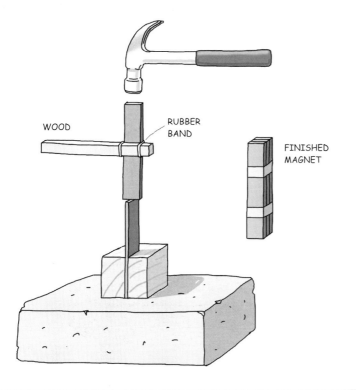

WOOD

RUBBER BAND

FINISHED MAGNET

MAKE A COMPASS

One of the reasons Scoresby was interested in making magnets was because with a magnet you can make a compass, and using a simple method like this, even shipwrecked sailors could find their way home.

You will need
- A needle, log pin or nail
- Magnet
- Polystyrene for a float
- A cup or glass

What to do
1. Stroke one end of the needle (e.g. from the centre to the point) ten times with one pole of the magnet.
2. Make a polystyrene float about as wide as the length of the needle. One very easy way of doing this is to cut the bottom out of an ordinary polystyrene cup.
3. Attach the needle to the float with Blu-Tack or tape.
4. Float the assembly in a cup of water. The needle should swing round and always point in the same direction – either north or south, according to which pole of the magnet you used.
5. Using a compass like this you should never get lost, even in thick fog on a dark moor, because you can keep walking in a straight line, and if you have some idea which direction is best you can easily stick to it.

WHY IT WORKS... The Earth is a giant magnet, with magnetic poles close to the geographic poles. That is why a compass needle points north, because it is attracted by the magnetic north pole.

When you live closer to the pole than to the equator – as we do in the UK – the lines of magnetic force dip down sharply into the Earth; so although the compass needle points horizontally north, in fact the magnetic field is stronger in a vertical direction. This is why Scoresby hammered his iron vertically.

When you put a piece of iron in a magnetic field, the iron concentrates the magnetism. The magnetic field is stronger in the iron than in the air around it, because of the particular arrangement of the electrons in each atom of iron.

When you bash the iron with a hammer, you shake up the atoms in it, and some of them realign themselves in such a way as to intensify the magnetic field. With each bash of the hammer, the iron becomes slightly more magnetized.

When you bash a piece of iron with another piece that is already magnetized, you amplify the effect. Because the first piece is magnetized, the second piece experiences a stronger magnetic field than the Earth's field, and so each time you bash it, more atoms are realigned, and the iron is magnetized more strongly. This means that the more you bash, the more the iron is magnetized.

WATER CLOCK

When defending themselves in court, the ancient Greeks had a limited period of time in which to state their case. A *clepsydra*, a jar with a small hole in the bottom, was filled with water, and the defendant was allowed to speak until the water had all run out. A cunning inventor called Ctesibios turned the *clepsydra* into a clock, and you can make one too.

BEWARE: WATER IS BOUND TO GET SPILLED. WE ADVISE YOU TO USE YOUR CLOCK OUTSIDE, OR ELSE IN THE KITCHEN SINK OR THE BATH!

You will need
- A sharp tool, e.g. a nail or pair of compasses
- 2 old plastic bottles, preferably tall and thin (1-litre plastic milk bottles are ideal)
- Approx 10cm plastic tube
- Waterproof glue
- A stand, or some method of holding one bottle above the other
- Pen and paper for marking measurements

As you have to fit one bottle on top of another under a tap, you may not have enough room. If necessary, use a shorter container for the top one, e.g. a yogurt pot. The taller the lower container is, the more precise your measurements will be.

What to do
1. Use your sharp tool to make a small hole in the bottom of one of the bottles. You could fill the bottle with water and then let it trickle out through the hole; you now have a *clepsydra*.
2. In the same bottle, drill a hole near the top of one of the sides – a hole the same size as your plastic tube.
3. Glue the tube into the hole so that almost all its length is outside, and tilting slightly downwards.

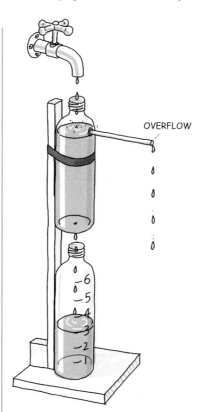

OVERFLOW

4. Fix this bottle above the other, and under a tap. Holding one finger over the hole in the bottom, fill the bottle until it begins to overflow from your side tube.
5. Leave the tap dripping, so that the water keeps overflowing and the water level in the top bottle always stays the same. This was Ctesibios's brilliant idea. A constant depth (or 'head') of water in the top bottle ensures a steady flow through the hole in the bottom.

6. When you're ready, take your finger away from the hole, and let the water drip into the bottom bottle.

Calibration
1. After one minute, mark the water level in the bottom bottle. Either use a pen that will write on plastic, or tape a strip of paper up the outside.
2. Mark the level after two minutes, three minutes, and so on.
3. Once you have calibrated your clock, you can empty out the bottom bottle and use it to time events.
4. The faster the flow through your hole, the more quickly the bottom bottle will fill. Make several holes – better than one big hole – and the bottom bottle will fill in, say, five minutes. This will provide you with an accurate clock for measuring, say, two or three minutes. However, if you would rather be able to measure an hour, then you will need a much smaller hole.
5. You might even like to make two clocks side by side, feeding the second with the overflow from the first. Then you could have one to measure minutes and the second to measure hours.

More things to try
● Use your kitchen balance for calibration and measurement.

MAKE SURE THE BALANCE WILL NOT BE RUINED BY GETTING WET!

Put the lower container (which should be low and squat with a wide mouth) on the balance pan, and note the weight shown after each minute.

● Get your clock to sound an alarm, so that you can use it as a timer. There are various ways to do this, but you might like to try putting a float (say polystyrene or cork) in the bottom container, and arranging for it to push against a balanced can, or some other trigger, when water has run in for, say, five minutes.

When the float pushes it, the can falls off into the sink.

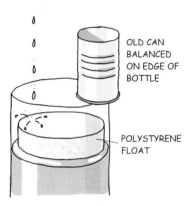

OLD CAN BALANCED ON EDGE OF BOTTLE

POLYSTYRENE FLOAT

DID YOU KNOW?
Ctesibios lived in the intellectual Greek community at Alexandria in Egypt around 250 BC, and may have been a pupil of Hero of Alexandria (see page 20). He was a barber, and he invented not only the water clock and a counter-balanced barber's mirror, but also the water organ and the force pump – so he must have been a remarkable man.

MINUTE-HAND CLOCK

RATING: ✳ EASY!

Ctesibios was keen on alarms, and used to fit his clocks with all sorts of amusing devices, including singing birds. Surprisingly, however, he seems not to have bothered to try to make the measurement of time more precise; he was content to read it to the nearest hour. However, you might like to try and improve on his rather crude system and build a clock with a minute hand.

You will need
● Tools and equipment as in the demonstration on the opposite page
● Piece of string
● A cork or polystyrene float
● Screw eye or tape
● Cotton reel (empty) for the pulley
● A nail
● Small weight
● Cardboard

What to do
1. Attach a piece of string to the cork or polystyrene float, using a screw eye or sticky tape.
2. Pass the string over a pulley – the cotton reel is good for this – which is free to rotate on a nail in the wooden upright. Tie a small weight to the other end of the string.
3. Make a minute hand from cardboard or flyweight envelope stiffener and tape or glue it to the cotton reel.
4. Make a clock face from cardboard and fix it to the wooden stand behind

the cotton reel. Don't write numbers on the clock face in advance. Instead, mark a zero at the bottom left (about 7 o'clock!).
5. Let the clock run by taking your finger off the hole in the top bottle as before, and mark each minute when you see where the hand reaches. This should give you a timer much more precise than those made by Ctesibios.

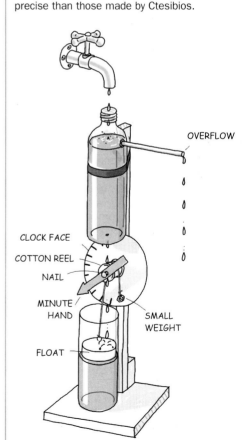

OVERFLOW

CLOCK FACE
COTTON REEL
NAIL
MINUTE HAND
FLOAT
SMALL WEIGHT

PENDULUM TIMER

In 1583, the 19-year-old Galileo was bored by a long sermon in the cathedral at Pisa, and began to watch the bronze lamp swinging. He measured the time of each swing by feeling his pulse and counting the number of heartbeats. He noticed that the lamp always took exactly the same time for each swing, whether it was a small swing or a large one.

You will need
- A reel of cotton
- A small weight, such as a metal nut or a £1 coin
- Sticky tape
- A metal hook or eye to hang your pendulum
- A watch or clock with a second hand – or you too could use your pulse!
- For the support, you could make a wooden stand, or just fasten a piece of wire to the back of a chair or the edge of a table

What to do
1. Break off about a metre of cotton thread. Fix your weight to one end of the thread to make the bob of the pendulum; tie the thread through the nut, or tape it to the coin.
2. Pass about 30cm of thread through the eye or hook, take a turn around the metal, and tape the thread to the support.
3. Measure the length of the thread between the eye and the bob, and write the length down.
4. Pull back the pendulum about 30 degrees and let go. Note how many seconds it takes to complete ten whole swings; one swing is away from you and back to where you let it go.
5. Repeat for ten more swings.
6. Write down the average time for ten swings, and divide by 10 to get the time for one swing.
7. You now have a precise timer. As long as you don't change anything,

your pendulum, like Galileo's lamp, will always take exactly the same time for one swing, as long as it does not swing more than about 30 degrees from the vertical.

More things to try
● Try varying the length of the thread. Untape the top and lengthen it by, say, 15cm. Repeat the test above, and see how long the swing takes now. How long a thread do you need for a swing of just one second? How long for just two seconds? Can you make a pendulum that takes four seconds for one swing?
● What is the connection between the length of the thread and the time of swing?
● Try getting the timer to make a

noise at some point during each swing. You could hang something light but hard – perhaps a pin – below your bob, and let it just hit a bell or empty tin can on each swing. This would give you an excellent timer for the Galileo experiments on page 96.

DID YOU KNOW?
Astronomers once used pendulums as precise timers, but in order to keep them swinging they had to pay children to push them every now and then. In the 1630s Galileo had the idea of using a pendulum in a clock, but by then he was blind, and it was Christiaan Huygens who made the first pendulum clock after Galileo died (see page 74).

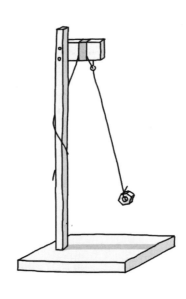
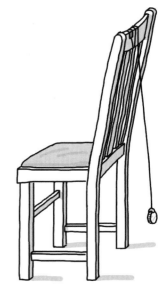

THE AUTHORS

Adam Hart-Davis is a freelance photographer, writer and broadcaster, probably best known as the mountain-biking, fluorescently clad presenter of the BBC2 series *Local Heroes*. He has also presented two history series on BBC2 and one on Carlton, and a BBC Radio 4 series called *Inventors Imperfect*.

Before he began presenting, Adam spent five years in publishing and 17 years at Yorkshire Television, first as a researcher and then as producer of such series as *Scientific Eye* and *Arthur C. Clarke's World of Strange Powers*.

He has written around 12 books including *World's Weirdest 'True' Ghost Stories*, *Test Your Psychic Powers* (with Susan Blackmore), *Thunder, Flush & Thomas Crapper (an EncycLOOpedia)* and *Eurekaaargh!* (about inventions that nearly worked). He has also written two *Local Heroes* books with Paul Bader.

Adam is a member of the Bureau of Freelance Photographers, the Newcomen Society and the British Toilet Association, and a Fellow of the Royal Society of Arts. He won the 1999 Gerald Frewer memorial trophy of the Council of Engineering Designers.

Paul Bader intended to become a scientist, studying genetics at Edinburgh University, before realizing that what he really wanted to do was to put science on television. At Yorkshire Television, from 1982, he worked on many science and medicine series for the ITV network with some great presenters – Miriam Stoppard, David Bellamy and Rob Buckman. A fellow producer was soon-to-be cyclist, Adam Hart-Davis.

Paul left Yorkshire Television in 1991 to set up on his own as Screenhouse Productions, and has now made six series of *Local Heroes* for the BBC, as well as nearly 80 history programmes.

He lives in Leeds with his wife and daughters, who for some years now have been used to strange things being made in the basement and wooden fax machines occupying the dining room.

Paul Bader has also written two *Local Heroes* books with Adam Hart-Davis. He was winner of the 1999 Chemical Industry Association award for his contribution to the public understanding of science.

INDEX